OPTIMIZATION IN ECONOMICS AND FINANCE

Dynamic Modeling and Econometrics in Economics and Finance

VOLUME 7

Series Editors

Stefan Mittnik, *University of Kiel, Germany*
Willi Semmler, *University of Bielefeld, Germany* and
New School for Social Research, U.S.A.

The titles published in this series are listed at the end of this volume.

Optimization in Economics and Finance

Some Advances in Non-Linear, Dynamic, Multi-Criteria and Stochastic Models

by

BRUCE D. CRAVEN
University of Melbourne, VIC, Australia

and

SARDAR M. N. ISLAM
Victoria University, Melbourne, VIC, Australia

Springer

A C.I.P. Catalogue record for this book is available from the Library of Congress.

ISBN 978-1-4419-3714-8 (PB)
ISBN 978-0-387-24280-4 (e-book)

Published by Springer,
P.O. Box 17, 3300 AA Dordrecht, The Netherlands.

Sold and distributed in North, Central and South America
by Springer,
101 Philip Drive, Norwell, MA 02061, U.S.A.

In all other countries, sold and distributed
by Springer,
P.O. Box 322, 3300 AH Dordrecht, The Netherlands.

Printed on acid-free paper

Table of Contents

Preface

Many optimization questions arise in economics and finance; an important example of this is the society's choice of the optimum state of the economy (which we call a social choice problem). This book,

Optimization in Economics and Finance,

extends and improves the usual optimization techniques, in a form that may be adopted for modelling optimal social choice problems, and other related applicastions discussed in section 1.2, concerning new[3] economics. These types of optimization models, based on welfare economics, are appropriate, since they allow an explicit incorporation of social value judgments and the characteristics of the underlying socio-economic organization in economic and finance models, and provide realistic welfare maximizing optimal resource allocation and social choices, and decisions consistent with the reality of the economy under study. The methodological questions discussed include:

- when is an optimum reached, and when is it unique?
- relaxation of the conventional convex (or concave) assumptions on an economic or financial model,
- associated mathematical concepts such as *invex* (relaxing *convex*) and *quasimax* (relaxing *maximum*),
- multiobjective optimal control models, and
- related computational methods and programs.

These techniques are applied to models of economic gropwth and development, including

- small stochastic perturbations,
- finance and financial investment models (and the interaction between financial and production variables),
- modelling sustainability over long time horizons,
- boundary (transversality) conditions, and
- models with several conflicting objectives.

Although the applications are general and illustrative, the models in this book provide examples of possible models for a society's social choice for an allocation that maximizes welfare and utilization of resources. As well as using existing computer programs for optimization of models, a new computer program, named SCOM, is presented in this book for computing social choice models by optimal control.

This book contains material both unpuhlished and previously published by the authors, now rearranged in a unified framework, to show the relations between the topics and methods, and their applicability to questions of social chnoice and decision making.

This book provides a rigorous study on the interfaces between mathematics, computer programming, finance and economics. The book is suitable as a reference book for researchers, academics, and doctoral students in the area of mathematics, finance, and economics.

The models and methods presented in this book will have academic and profesional application to a wide range of areas in economics, finance, and applied mathematics, including optimal social choice and policy planning, use of optimal models for forecasting, market simulation, developmenjt planning, and sensitivity analysis.

Since this is an interdisciplinary study involving mathematics, economics, finance and computer programming, readers of this book are expected to have some familiarity with the following subjects: Mathematical Analysis, Optimal Control, Mathematical Finance, Mathematical Economics, Mathematical Programming, Growth Economics, Economic Planning, Environmental Economics, Economics of Uncertainty, Welfare Economics, and Computational Economics.

The various SCOM computer programs listed in this book may also be downloaded from the web site: *http://bdc.customer.netspace.net.au* .

The authors thank Margarita Kumnick for valuable proof-reading and checking. The authors also thank the Publishing Editor of Kluwer, Mrs Cathelijne van Herwaarden, and a referee, for their cooperation and support in the completion of this book.

B. D. Craven
Dept. of Mathematics
& Statistics
University of Melbourne
Australia

S. M. N. Islam
Centre for Strategic
Economic Studies
Victoria University, Melbourne
Australia

1 September 2004

The authors

Dr. B. D. Craven was (until retirement) a Reader in Mathematics at University of Melbourne, Australia, where he taught Mathematics and various topics in Operations Research for over 35 years. He holds a D.Sc. degree from University of Melbourne. His research interests include continuous optimization, nonlinear and multiobjective optimization, and optimal control and their applications. He has published five books, including two on mathematical programming and optimal control, and many papers in international journals. He is a member of Australian Society for Operations Research and INFORMS.

Prof. Sardar M. N. Islam is Professor of Welfare and Environmental Economics at Victoria University, Australia. He is also associated with the Financial Modelling Program, and the Law and Economics Program there. He has published 11 books and monographs and more than 150 technical papers in Economics (Mathematical Economics, Applied Welfare Economics, Optimal Growth), Corporate Governance, Finance, and E-Commerce.

Acknowledgements and Sources of Materials

The authors acknowledge permission given by the following publishers to reproduce in this book some material based on their published articles and chapters:

Chapter 3 is based (with some additional material) on *Computing Optimal Control on MATLAB - The SCOM Package and Economic Growth Models,* Chapter 5 in *Optimisation and Related Topics,* Eds, A. Rubinov et al. Volume 47 in the Series *Applied Optimization,* Kluwer Academic Publishers, 2001. Kluwer has given permission to reproduce this article.

Chapter 4 is (with minor changes) the paper *Computation of Non-Linear Continuous Optimal Growth Models: Experiments with Optimal Control Algorithms and Computer Programs,* 2001, *Economic Modelling: The International Journal of Theoretical and Applied Papers on Economic Modelling,* Vol. 18, pp. 551-586, North Holland Publishing Co.

Chapter 6 is (with minor modifications) the paper *Measuring Sustainable Growth and Welfare: Computational Models, Methods,* Editors: Thomas Fetherston and Jonathan Batten, *"Governance and Social Responsibility"*, Elsevier-North Holland, 2003.

Chapter 9 is the paper *Transversality Conditions for Infinite Horizon Optimal Control Models in Economics and Finance*, submitted to Journal of Economic Dynamcs and Control, an Elsevier publication.

Elsevier have stated that an Elsevier author retains the right to use an article in a printed compilation of works of the authors.

The book also includes excerpts from other papers by the authors, often rearranged to show relevance and relation to other topics; these items are acknowledged where they occur. The material in this book is now presented in a unified manner, with a focus on applicability to economic issues of social choice.

Chapter 1

Introduction : Optimal Models for Economics and Finance

1.1. Introduction

This book is concerned with applied quantitative welfare economics, and describes methods for specification, analysis, optimization and computation for economic and financial models, capable of addressing normative social choice and policy formulation problems. Here *social choice* refers to the optimal intertemporal allocation of aggregate and disaggregate resources. The institutional and organizational aspects of achieving such allocation in a society are not discussed here. Zahedi (2001) has surveyed other methods for social choice. The book aims to provide some extensions and improvements to the traditional methods of optimization, as applied to economics and finance, which could be adopted for social decision making (*social choice*) and related applications. The mathematical techniques include nonlinear programming, optimal control, stochastic modelling, and multicriteria optimization.

Many questions of optimization and optimal control arise in economics and finance. An optimum (maximum or minimum) is sought for some objective function, subject to constraints (equalities or inequalities) on the values of the variables. The functions describing the system are often nonlinear. For a time-dependent system, the variables become functions of time, and this leads to an optimal control problem. A control function describes a quantity (such as consumption, or investment) that can be controlled, within some bounds. A state function (such as capital accumulation) takes values determined by the control function(s) and the dynamic equation(s) of the system.

Some recent developments in the mathematics of optimization, including the concepts of invexity and quasimax, have not previously been applied to models of economic growth, and to finance and investment. Their applications to these areas are shown in this book. Some results are presented concerning when an optimal control model has a unique optimum, what happens when the usual convexity assumptions are weakened or absent, and stability to small disturbances of the model or its parameters. A new computational package called SCOM, for solving optimal control problems on MATLAB, is introduced. It facilitates computational experiments, in which there are changes to model features or parameters.

These developments are applied, in particular, to:

- models of optimal (welfare maximizing) intertemporal allocation of resources.

- economic growth models with a small stochastic perturbation.
- models for finance and investment, including some stochastic elements, and especially considering the interaction between financial and production variables.
- modelling sustainability over a long (perhaps infinite) time horizon.
- models with several conflicting objectives.
- boundary (*transversality*) conditions.

These extended results can be usefully applied to various questions in economics and finance, including social decision making and policy analysis, forecasting, market simulation, sensitivity analysis, comparative static and dynamic analysis, planning, mechanism design, and empirical investigations. If an economic system behaves so as to optimize some objective, then a computed optimum of a model may be used for forecasting some way into the future. However, the book is focussed on optimal social decision making (social choice).

1.2. Welfare economics and social choice: Modelling and Applications

A central issue in economics and finance, concerning welfare economics, is to find a normative framework and methodology for social decision-making, so as to choose the socially desirable (multi-agent or even aggregate) state of the economy, a task popularly known as "social choice". Optimisation methods based on welfare economics can aid such social decision making ((Islam 2001a).

The optimisation models of economics and finance can, therefore, be interpreted as *models for normative social choice* which specify optimal social welfare in the economy and financial sector satisfying the static and dynamic constraints of the economy since these models can generate a set of aggregative and disaggregstive optimal decisions. choices or allocation of resources for the society. This approach is in the line of arguments advanced in the paradigm of new^3 welfare economics (Islam 2001b; Clarke and Islam, 2004). It has the following main elements: 1) the possibility perspective of social choice theory; 2) measurability of social welfare based on subjective or objective measures; 3) the extended welfare criteria; 4) operationalisation of welfare economics and social choice (which was the original motivation of classical economists for developing the discipline of welfare economics), and 5) a multi-disciplinary system approach incorporating welfaristic and non-welfaristic elements of social welfare.

Any welfare economic analysis of issues in economics and financial policies involves the application of the following multidisciplinary criteria of moral philosophy and welfare economics: efficiency, rationality, equity, liberty, freedom, capabilities and functioning (see Hausman and McPherson, 1996) for a survey of these criteria). This framework of new^3 welfare economies provides the scope for evaluating economic outcomes in terms of social welfare (and efficiency, utility) as well as other criteria of welfare economics and moral philosophy such as rights, liberty, morality, etc. (see Hausman and McPherson, 1996 for a survey of the concepts and issues and their economic implications).

The incorporation of this approach in optimisation modelling is possible through the choice of the social discount rate, the objective function (extended welfare criteria incorporating welfaristic and non-welfaristic elements of social welfare), terminal conditions, time horizon, and the modelling structure.

In making such an application of optimisation models, several conceptual and methodological issues in social choice theory and welfare economics (which have dominated the controversy about the possibility of social choice) needs to be resolved including the following (Islam 2001b):

- The nature value judgment about the nature of individual well-being or welfare (such as in utilitarianism or welfarism, capability,) etc.
- Possibilities for measurability of utility and welfare (cardinality or ordinality).
- Interpersonal comparability of utility and welfare.
- The nature of marginal utility of income (constancy or variability).
- The role of distributional concerns in welfare judgment (the intensity of preferences).
- The choice of a measurement and accounting method (nature of preference indexing, numerical calculations, etc.).
- The extent of informational requirements for decision making.

These issues can be considered from the impossibility (Arrow, 1951) or possibility perspectives (Sen, 1970). The possibility perspective approach requires a set of axioms including cardinality, intertemporal comparability, and the relevance of the intensity of preferences. In this possibility approach (see Sen 1999), there is an urge for the need for, amongst others, finding a suitable method and information broadening for developing an optimistic social choice theory for useful social welfare analysis and judgment. This can be accomplished by developing an operational approach to social choice. This is an especially immediate task in applied welfare economics, although work in this area has not progressed far. In Islam (2001a, 2001b) and Clarke and Islam (2004), a paradigm has been developed for new^3 welfare economics, for normative operational social choices based on the possibility perspective.

The choice of the elements for a social norm is controversial, since each specification relates to some form of value judgment in a welfare economics choice model, and a choice significantly affects the pattern and level of social welfare. A specification of the the elements of a social choice should be based within the framework of some paradigm of welfare economics. The new^3 welfare economics paradigm adopts the following set of assumptions and elements of an operational approach to social choice aqnd decision making:

- Definition of well-being and welfare: the social welfaristic approach (Islam 2001a and 2001b) - economic activities, which improve net social welfare, are justified.
- The possibility of the specification of aggregate social welfare criteria and index: the possibility theorem perspective.
- Time preference: different discounting approaches for intertemporal

equity - depending on the preference of the society.

- Units of measurement; market and shadow prices of goods and services
- Methods for modelling: efficient allocation or optimisation modelling.
- Institutions: various alternative institutions can be assumed such as competitive market economy, mixed economy, or planning - depending on the underlying social organization.

The main argument of this book is that mathematical models can be developed, incorporating the above elements of new[3] welfare economics; they can provide useful information to understand social choice in relevant economic, social, environmental and financial issues, and formulating appropriate policies.

The general structure of an optimisation model in economics and finance, containing the above elements, and suitable for normative social choice or decision making (see Craven 1995; Islam 2001a; Laffont, 1988) is as follows:

$$W = f(y) \text{ subject to } g(y) \in S,$$

where: $I = [a, b]$;

W is an indicator of social welfare;

V is the space of functions;

$f(y)$ is a scalar or vector valued social welfare functional of society.

y is a vector of variables or functions of economic and financial sub-systems;

$g(y)$ is a constraint function (including economic and financial factors);

S is a convex cone, describing a feasible set of the economy; and

$\mathbf{R}^n$ is Euclidian space of n dimensions.

In the above social welfare model, a social welfare function of the Bergson-Samuelson form is specified to embed social welfare judgments about alternative states of resource allocation in the economy. (For further details, see Islam, 2001a.) Social welfare, and factors affecting it, are assumed to be measurable and quantifiable. The problem of normative social choice in decision making is represented by the optimization model, based on the possibility perspective of social choice. It is operational, since it may be applied to real life conditions, for finding optimal decisions in society. The model can represent the economic organization of a competitive market or planning system (the selection of a system of social organization depends on the social preferences assumed in the model). A model, containing an objective function, constraints and boundary conditions, can represent the socio-economic factors relevant for decision making. *These general assumptions are made for the various models in this book*; specific assumptions for each model are discussed in the relevant cases.

The optimal solution to the welfare optimisation social choice problem exists (i.e., an optimal decision, choice, or policy exits) if the problem satisfies the Weierstrass theorem; and if the objective function is convex, x* is a global solution.

The set S represents the static or dynamic economic and financial systems. The objective function f(x) is the social welfare functional embedding social choice criteria. Different value judgements and different theories of welfare economics and social choice, and various sub-systems of the economy can be incorporated in this social choice program by making different assumptions about different functions, parameters and the structure of the above model. The above control model can embed and address the issues of welfare economics and social choice discussed above if it is based on a proper specification of the method for aggregation of individual welfare, welfare criteria, cost benefit consideration, and institutional mechanisms assumed for society. *The results of the model can specify the optimal choices* regarding optimal dynamic welfare and resource allocation and price structure, the optimal rate and valuation of consumption, capital accumulation, and other economic activities, and optimal institutional and mechanism design. Further discussion on construction of welfare economic modelling is given in Islam (2001a), Heal (1973), Chakravarty (1969), and Fox, Sengupta and Thorbecke (1973).

The above social choice model is a finite horizon free terminal time continuous optimisation problem and it is deterministic, and open loop with social welfare maximization criteria. Other possible forms of social choice models include dynamic game models and with other types of end points and transversality conditions; overtaking, catching up and Rawlsian optimality criteria; with different types of constraints, discontinuities and jumps; and with uncertainty. These social choice models may also represent equilibrium and disequilibrium economic systems, adaptive dynamics, social learning, chaotic behaviour, artificial intelligence and genetic algorithm.

In such an optimisation model of social choice, the following set of elements should be specified:

- an economic model (including social, financial, and environmental constraints;
- the length of the planning horizon;
- the choice of an optimality criterion or an intertemporal utility function;
- the discount rate, representing the rate of time preference; and
- the terminal or transversality conditions.

The specification of the elements is a political economic exercise involving substantial value judgment on the part of the modeller. Depending on the value judgment of the modeller, a particular form of each element can be specified.

1.3. The objectives

The objective of this book is to provide extensions to the existing methods for optimisation in economics and finance which can be appropriately used for normative social choice, based on the possibility perspective of social choice and the other elements of new[3] welfare economics discussed above, as well as for other exercises such as sensitivity analysis, simulation of market behaviour, forecasting, and comparative static and dynamic analysis. The focus of the book is on the methods for optimisation, not on the social choice issues in optimisation models, in economics and finance. This book has not taken any particular perspective in social value judgments, and therefore the details of the choice of the elements are not provided here. We have left the specification of various elements of welfare economics and optimisation modelling in a possible general form. A modeller can choose a set of specific elements according to his/her value judgment (see also Islam, 2001a), to develop a model for a particular economy.

Although a variety of models and computation approaches are developed and implemented in this book, *they may all describe social choices,* concerning the maximization of social welfare, intertemporal allocation, and utilization of resources, in relation to the social value judgement expressed in the models.

1.4. An example of an optimal control model

A large part of this book is concerned with optimal control models for economic questions. Such models are generally of the form:

$$\mathrm{MAX}_{x(.),u(.)}\ F^0(x,u) := \int_0^1 f(x(t),u(t),t)dt + \Phi(x(1))$$

$$\text{subject to } x(0) = a,\ \dot{x}(r) = m(x(t),u(t),t),\ q(t) \leq u(t) \leq r(t) \quad 0 \leq t \leq 1).$$

Here the *state function* $x(t)$ could describe capital, the *control function* $u(t)$ could describe consumption; an objective (an integral over a time period, plus an endpoint term) describes a utility to be maximized, subject to a *dynamic equation*, a differential equation determining the state.

A special case is a model for economic growth and development, of which the following is an example. The well known Kendrick-Taylor model for economic growth (Kendrick and Taylor, 1971) describes the change of capital stock $k(t)$ and consumption $c(t)$ with time t by a dynamic differential equation for the time derivative $\dot{k}(t)$, and seeks to maximize a discounted utility function of consumption, integrated over a time period $[0,T]$. The model is expressed as:

$$\mathrm{MAX} \int_0^T e^{-\rho t} c(t)^\tau dt \quad \text{subject to } k(0) = k_0,$$

$$\dot{k}(t) = \zeta e^{qt} k(t)^\beta - \sigma k(t) - c(t), \quad k(T) = k_T.$$

No explicit bounds are stated for $k(t)$ and $c(t)$. However, both the formulas and their interpretation requires that both $k(t)$ and $c(t)$ remain positive. However, with some values of $u(t)$, the differential equation for $k(t)$ can bring $k(t)$ down to zero. The capital is the *state function* of this optimal control formulation, and the consumption is the *control function.* In general the control function is to be varied, subject to any stated bounds, in order to achieve the maximum. This model includes the standard features, namely an optimality criterion contained in an objective function which consists of the discounted sums of the utilities provided by consumption at every period, a finite planning horizon T, a positive discount rate, boundary conditions, namely initial values of the variables, and parameters and the terminal conditions on the state.

1.5. The structure of the book

Chapter 2 presents the relevant mathematics of optimization, and especially optimal control, including the formulation of dynamic economic and finance models as optimal control problems. Questions discussed include the following:

- When is an optimum reached, and when is it unique?
- Relaxing of convex assumptions, and of maximum to quasimax.
- Multiobjective optimal control, and the Pontryagin conditions for optimality for single-objective and multiobjective problems.

Some qualitatively different effects may occur with nonconvex models, such as non-unique optima, and jumps in the consumption function, which have economic significance.

In Chapter 3, algorithms for computing optimal control are discussed, with reasons for preferring a direct optimization approach, and step- function approximations. A computer package *SCOM* is described, developed by the present authors, for solving a class of optimal control problems in continuous time, using the MATLAB system, but in a different way from the *RIOTS_95* package (Schwartz, 1996), which also uses MATLAB. As in the MISER (Jennings et al., 1998) and OCIM (Craven et al., 1998) packages, the control is parametrised as a step-function, and MATLAB's *constr* package for constrained optimization is used as a subroutine. End-point conditions are simply handled using penalty terms. Much programming is made unnecessary by the matrix features built into MATLAB. Some economic models present computational difficulties because of implicit constraints, and there is some advantage using finite difference approximations for gradients. The Kendrick-Taylor model of economic growth is computed as an example.

Chapter 4 discusses the use of optimal control methods for computing some non-linear continuous optimal welfare, development, and growth models. Results are reported for computing the Kendrick-Taylor optimal-growth model using RIOTS_95 and SCOM programs based on the discretisation approach. Comparisons are made to the computational experiments with OCIM, and MISER. The results are used to compare and evaluate mathematical and economic properties, and computing criteria. While several computer packages are available for optimal control problems, they are not always suitable for particular classes of control problems, including some economic growth models.

Chapter 5 presents some proposed extensions for dynamic optimization modelling in finance, for characterizing optimal intertemporal allocation of financial and physical resources, adapted from developments in other areas of economics and mathematics. The extensions discussed concern (a) the elements of a dynamic optimization model, (b) an improved model including physical capital, (c) some computational experiments. It is sought to model, although approximately, the interaction between financial and production variables. Some computed results from simulations are presented and discussed; much more remains to be done.

Chapter 6 develops mathematical models and computational methods for formulating sustainable development and social welfare programs, and discusses approaches to computing the models. Computer experiments on modifications of the Kendrick-Taylor growth model, using the optimal control packages SCOM (Craven & Islam, 2001) and RIOTS_95 (Schwartz 1989), analyse the effects of changing the discount factor, time scale, and growth factor. These packages enable an economist to experiment, using his own computer, on the results of changing parameters and model details.

Chapter 7 presents a non-linear optimal welfare, development, and growth model under uncertainty, when the stochastic elements are not too large. Methods of describing the stochastic aspect of a growth model are reviewed, and computational and growth implications are analysed. The Kendrick-Taylor model is modified to a stochastic optimal control problem, and results are computed with various parameters. The model results have implications concerning the structure of optimal growth, resource allocation, and welfare under uncertainty. They show that the stochastic growth can be modelled fairly simply, if the variance is small enough not to dominate the deterministic terms.

Chapter 8 discusses a number of welfare models, both for static models (not time-dependent) and for dynamic models (evolving in time). The models include welfare models where each user gives some weight to the welfare of other users, cooperative game models, and several multiobjective optimal control models, for resource allocation, development, growth, and planning. Questions of stability to perturbation are discussed, also computational approaches.

Chapter 9 extends the existing literature on transversality conditions for infinite-horizon optimal control models of social choice in economics and finance. In optimal control models with infinite horizon in economics and finance, the role and validity of the boundary condition for the costate function (called the transversality condition) has been much discussed. This chapter derives such conditions, and proves their validity, under various assumptions, including the cases: (i) where the state and control functions tend to limits ("steady state"), and some gradient conditions hold, (ii) when the state and control tend sufficiently rapidly to limits, and (iii) where there is no steady state, but the model may be normalized to allow for a growth rate. Shadow price interpretations are discussed, also sufficient conditions for optimality. A nonlinear time transformation, and a normalization of the state and control functions, are used to convert the problem to a standard optimal control problem on a finite time interval. As well as establishing transversality conditions, this approach gives a computational method for infinite-horizon models of optimal social choice and decision making n economics and finance.

Chapter 10 presents the conclusions and findings of this research project.

Chapter 2
Mathematics of Optimal Control

2.1. Optimization and optimal control models[1]

This chapter discusses mathematical ideas and techniques relevant to optimization questions in economics and related areas, and particularly relevant for construction and application of social choice models, based on the assumptions of new^3 economics discussed in chapter 1. First considered is a *static* model, optimising over a vector variable z, typically with a finite number of components. When the variable z is not static, but describes some variation over time, an optimal control model may be required, where the objective is typically an integral over a time horizon, say $[0, T]$), with perhaps an additional term at the final time T, and the evolution over time is described by a dynamic equation, typically a differential equation. This leads to an optimal control model, where z becomes a function of time t. In each case, a minimum may be described by necessary *Karush-Kuhn-Tucker (KKT) conditions*, involving Lagrange multipliers. When the time t is a continuous variable, a related set of necessary *Pontryagin conditions* often apply (see section 2.2).

There are also discrete-time models, where the integration is replaced by summation over a discrete time variable, say $t = 0, 1, 2, ..., T$, and the dynamic equation is a difference equation. For discrete time, the KKT conditions apply, but not all the Pontryagin theory.

Questions arise of *existence* (thus, when is a maximum or minimum reached?), *uniqueness* (when is there exactly one optimum?), relaxation of the usual assumption of convex functions, and what happens to a dual problem (in which the variables are the Lagrange multipliers) in the absence of convex assumptions? These are discussed in sections 2.3 through 2.7. Further issues arise when there are several conflicting objectives; these are discussed in sections 2.8 through 2.14.

Consider first a mathematical programming model (e.g a model for a normative social choice problem in economics and finance):

$$\text{MIN } f(z) \text{ subject to } g(x) \leq 0, k(z) = 0,$$

in which an objective function $f(\cdot)$ is maximized, with the state variable z constrained by inequality and equality constraints. The functions f, g and h are assumed differentiable. Note that a maximization problem, MAX $f(z)$, may be considered as minimization by MIN $-f(z)$. Assume that a local minimum is reached at a point $z = p$ (*local* means that $f(z)$ reaches a minimum in some region around p, but not necessarily over all values of z satisfying the

[1] See also sections 4.2, 4.5, 6.3, 9.4.

constraints.) Assume that z has n components, $g(z)$ has m components, and $k(z)$ has r components. The gradients $f'(p),\ g'(p),\ k'(p)$ are respectively $1 \times n,\ m \times n,\ r \times n$ matrices.

Define the *Lagrangian* $L(z) := f(a) + \rho g(z) + \sigma k(z)$. The Karush-Kuhn-Tucker necessary conditions (*KKT*):

$$L'(p) = 0,\ \rho \geq 0,\ \rho g(p) = 0$$

then hold at the minimum point p, for some Lagrange multipliers ρ and σ, provided that some *constraint qualification* holds, to ensure that the boundary of the *feasible region* (satisfying the constraints) does not behave too badly). The multipliers are written as row vectors, with respectively m and r components. These necessary KKT conditions are *not* generally sufficient for a minimum. In order for (KKT) at a feasible point p to imply a minimum, some further requirement on the functions must be fulfilled. It is enough if f and g are convex functions, and k is linear. Less restrictively, *invex* functions may be assumed - see section 2.4.

Consider now an optimal control problem, of the form:

$$\text{MIN}_{x(.),u(.)}\ F^0(x,u) := \int_0^1 f(x(t),u(t),t)dt + \Phi(x(1))$$

subject to $x(0) = a,\ \dot{x}(r) = m(x(t),u(t),t),\ q(t) \leq u(t) \leq r(t)\ \ 0 \leq t \leq 1)$.

(This problem can represent the problem of optimizimg the intertemporal welfare in an economics or finance model.) Here the state function $x(.)$, assumed piecewise smooth, and the control function $u(.)$, assumed piecewise continuous, are, in general, vector-valued; the inequalities are pointwise. A substantial class of optimal control problems can (see Craven, 199/5); Craven, de Haas and Wettenhall, 1998) be put into this form; and, in many cases, the control function can be sufficiently approximated by a step-function. A terminal constraint $\sigma(x(1)) = b$ can be handled by replacing it by a penalty term added to $F^0(x,u)$; thus the objective becomes:

$$F(x,u) := F^0(x,u) + \tfrac{1}{2}\mu\|\sigma(x(1)) - b^*\|^2,$$

where μ is a positive parameter, and k approximates to b. In the augmented Lagrangian algorithm (see e.g. Craven, 1978), constraints are thus replaced by penalty terms; μ is finite, and typically need not be large; here $b^* = b + \theta/\mu$, where θ is a Lagrange multiplier. If there are few constraints (or one, as here), the problem may be considered as one of parametric optimization, varying b^*, without computing the multipliers. Here T is finite and fixed; the endpoint constraint $q(x(T)) = 0$ is not always present; constraints on the control $u(t)$ are not always explicitly stated, although an implicit constraint $u(t) \geq 0$ is commonly assumed. If $q(.)$ or $\Phi(.)$ are absent from the model, they are replaced by zero.

In a model for some economic or financial question of maximizing welfare, the state $x(.)$ commonly describes capital accumulation, and the control $u(.)$ commonly describes consumption. Both are often vector functions.

The differential equation, with initial condition, determines $x(.)$ from $u(.)$; denote this by $x(t) = Q(u)(t)$; then the objective becomes:

$$J(u) = F^0(Q(u), u) + \tfrac{1}{2}\mu\|\sigma(Q(u)(1)) - b^*\|^2,$$

Necessary Pontryagin conditions for a minimum of this model have been derived in many ways. In Craven (1995), the control problem is reformulated in mathematical programming form, in terms of a Lagrangian:

$$\int_0^T [e^{-\delta t} f(x(t), u(t)) + \lambda(t)m(x(t), u(t), t) - \lambda(t)\dot{x}(t) + \alpha(t)(q(t) - u(t))$$

$$+\beta(t)(u(t) - r(t) + \frac{1}{2}\mu[\Phi(x(t) - \mu^{-1}\rho]_+^2 + \frac{1}{2}\mu[q(x(T) - \mu^{-1}\nu]\delta(t-T)]\ dt.$$

with the costate $\lambda(t)$, and also $\alpha(t)$ and $\beta(t)$, representing Lagrange multipliers, μ a weighting constant, ρ and ν are Lagrange multipliers, and $\delta(t-T)$ is a Dirac delta-function. Here, the terminal constraint on the state, and the endpoint term $\Phi(x(T))$ in the objective, have been replaced by penalty cost terms in the integrand; the multipliers ρ and ν have meanings as shadow costs. (This has also computational significance – see section 3.1. The solution of a two-point boundary value problem, when $x(T)$ is constrained, has been replaced by a minimization.) The state and control functions must be in suitable spaces of functions. Often $u(.)$ is assumed piecewise continuous (thus, continuous, except for a finite number of jumps), and $x(.)$ is assumed piecewise smooth (the integral of a piecewise continuous function.)

The adjoint differential equation is obtained in the form:

$$-\dot{\lambda}(t) = e^{-\delta t} f_x(x(t), u(t)) + \lambda(t)m_x(x(t), u(t), t),$$

where f_x and m_x denote partial derivatives with respect to $x(t)$, together with a boundary condition (see Craven, 1995):

$$\lambda(T) = \Phi_x(x(T)) + \kappa q_x(x(T)),$$

in which Φ_x and q_x denote derivatives with respect to $x(T)$, and κ is a Lagrange multiplier, representing a shadow cost attached to the constraint $q(x(T)) = 0$. The value of κ is determined by the constraint that $q(x(T) = 0$. If $x(T)$ is *free*, thus with no terminal condition, and Φ is absent, then the boundary condition is $\lambda(T) = 0$. Note that $x(T)$ may be partly specified, e.g. by a linear constraint $\sigma^T x(T) = b$ (or $\geq$ b), describing perhaps an aggregated requirement for several kinds of capital. In that case, the terminal constraint differs from $\lambda(T) = 0$.

A diversity of terminal conditions for $\lambda(T)$ have been given in the economics literature (e.g. Sethi and Thompson, 2000); they are particular cases of the formula given above. For the constraint $q(x(T)) \geq 0$, the multiplier $\kappa \geq 0$.

From the standard theory, the gradient $J'(u)$ is given by:

$$J'(u)z = \int_0^1 (f + \lambda(t)m)_u(x(t), u(t), t)dt,$$

where the costate $\lambda(.)$ satisfies the adjoint differential equation:

$$-\dot{\lambda}(t) = (f + \lambda(t)m)_x(x(t), u(t), t), \ \ \lambda(1) = \mu(\sigma(x(1)) - b^*) + \Phi'(x(1));$$

$(.)_x$ denotes partial derivative. A constraint such as $\int_0^1 \theta(u(t))dt \leq 0$, which involves controls at different times, can be handled by adjoining an additional state component $y_0(.)$, satisfying $y_0(0) = 0, \ \dot{y}_0(t) = \theta(u(t))$, and imposing the state constraint $y_0(1) \leq 0$.The latter generates a penalty term $\frac{1}{2}\mu\|y_0(1)) - c\|_+^2$, where $c \approx 0$ and $[.]_+$ replaces negative components by zeros. (See an example of such a model in section 1.4).

2.2. Outline of the Pontryagin theory[2]

This section gives an outline derivation of the Pontryagin necessary conditions for an optimal control problem in continuous time, with a finite time horizon T. In particular, it is indicated how the boundary conditions for the costate arise; they are critical in applications. Comments at the end of this section i indicate what may happen with there is an *infinite time horizon* $(T = \infty)$. That discussion is continued in Chapter 9.

Consider now an optimal control problem:

$$\text{MIN } F(z, u) := \int_0^T f(x(t), u(t), t)dt \text{ subject to}$$

$$\dot{x}(t) = m(x(t), u(t), t), \ x(0) = x_0, \ g(u(t), t) \leq 0 \ \ (0 \leq t \leq T).$$

Here the vector variable z is replaced by a pair of functions, the *state function* (or *trajectory*) $x(\cdot)$ and the *control function* $u(\cdot)$. This problem can be written formally as:

$$\text{MIN } F(x, u) \text{ subject to } Dx = M(x, u), \ G(u) \leq 0.$$

Here D maps the trajectory $x(.)$ onto its gradient (thus the whole graph of $\dot{x}(t)$ $(0 \leq t \leq T)$. Consider the *Lagrangian* function :

$$L(x, u; \theta, \zeta) := F + \theta(-Dx + M) + \zeta G =$$

[2] See also sections 4.2, 5.7, 6.3.3, 6.5.2.

$$\int_0^T (f + \lambda m)dt - \int_0^T \lambda(t)\dot{x}(t)dt + \int_0^T \mu(t)g(u(t),t)dt,$$

in which the multiplier θ is represented by a *costate function* $\lambda(t)$, described by $\theta w = \int_0^T \lambda(t)w(t)dt$ for each continuous function w, and ζ is similarly represented by a function $\mu(.)$. Note that the *Hamiltonian* function:

$$h(x(t), u(t), t, \lambda(t)) := f(x(t), u(t), t) + \lambda(t)m(x(t)u(t), t)$$

occurs in the integrand. An integration by parts replaces the second integral by:

$$\lambda(T)x(T) - \lambda(0)x(0) + \int_0^T \dot{\lambda}(t)x(t),$$

in which $\lambda(0)x(0)$ may be disregarded, because of the initial condition on the state, $x(0) = x_0$.

If the control problem reaches a minimum, and certain regularity restrictions are satisfied, then necessary KKT conditions also hold for this problem, namely:

$$L_x = 0, \;\; L_u = 0, \zeta \geq 0, \;\; \zeta G = 0,$$

where suffixes x and u denote partial derivatives. The following is an outline of how the Pontryagin theory can be deduced, using (KKT). For a detailed account of this approach, especially including the (serious) assumptions required for its validity, see e.g. Craven (1995).

From $L_x = 0$ in (KKT), the *adjoint differential equation*:

$$-\dot{\lambda}(t) = h_x(x(t), \;\; \lambda(T)x(T) = 0$$

may be deduced, using the endpoint boundary condition $\lambda(T)x(T) = 0$ to eliminate the integrated part. The rest of (KKT) gives necessary conditions for minimization of the Hamiltonian with respect to the control only, subject to the constraints on the control. While they do not generally imply a minimum, they do in restrictive circumstances, leading to *Pontryagin's principle*, which states that the optimal control minimizes the Hamiltonian with respect to the control $u(t)$, subject to the given constraints on the control, while holding the state $x(t)$ and costate $\lambda(t)$ at their optimal values. The restrictions include the following:

- The control problem reaches a local minimum, with respect to the norm

$$\|u\|_1 := \int_0^T |u(t)|dt.$$

- The constraints on the control hold for each time t separately (so that constraints involving a combination of two or more times are excluded).
- Existence and boundedness of first and second derivatives of f and m.

The necessary Pontryagin conditions for a minimum of the control problem hence comprise:

- The dynamic equation for the state, with initial condition.
- The adjoint equation for the costate, with terminal condition.
- The Pontryagin principle.

If a terminal condition is omitted, then the system is not definitely defined, and generally uniqueness is lost. If the state has r components, then r terminal conditions are required. If $x^i(T)$ is not specified, then $\lambda^i(T) = 0$; if, however, $x^i(T)$ is fixed, then $\lambda^i(T)$ is free (not specified.)

This discussion has assumed a fixed finite time horizon T. If an infinite horizon is required, as in some economic models, then serious difficulties arise. It is not obvious that any minimum is reached. The objective $F(x, u)$ may be infinite, unless the function f includes a discount factor, such as $e^{-\delta t}$. The conditions on derivatives are not generally satisfied, over an infinite time domain, unless the state and control are assumed to converge sufficiently fast to limits as $t \to \infty$. The conjectured boundary condition $\lim_{t\to\infty}\lambda(t)x(t) = 0$ does not necessarily hold. Some circumstances where this boundary condition does hold are analysed in Chapter 9, with assumptions on convergence rates.

If the control problem is truncated to a finite planning interval $[0, T]$, with a terminal condition fixing $x(T)$ at the assumed optimal value for the infinite-horizon problem, then this gives the necessary conditions for the infinite-horizon problem, *except* that the terminal condition for the costate is omitted. So the system of conditions is not definitely defined, and often allows some additional, though spurious, solution. Various authors have adjoined a boundary condition (called *transversality condition* arbitrarily, to exclude the additional solutions. But it is preferable to obtain the correct boundary condition from a complete set of necessary conditions for a minimum (see Craven (2003) and Chapter 9.)

2.3. When is an optimum reached?

Questions arise of (i) existence of a maximum point $\mathbf{x}^*$, (ii) necessary conditions for a minimum, (iii) sufficient conditions for a maximum, (iv) uniqueness, (v) descriptions by *dual variables* (which interpret *Lagrange multipliers* as *prices*).

Consider first the maximization of an objective $f(x)$ over $x \in \mathbf{R}^n$. Concerning *existence*, if x is in $\mathbf{R}^n$, f and each g_i are continuous functions, and if the *feasible set* E of those x satisfying the constraints $g_i(x) \geq 0$ is *compact*, then at least one maximum point x* exists.

However, in an optimal control model in continuous time, the compactness property is usually not available. (If there are only a finite number of variables, then a set is compact if it is closed and bounded; but that does not hold for an infinite number of variables, as for example for a continuous state function.) Sometimes convex, or invex, assumptions can be used to show that an optimum is reached - see section 2.4.

Assuming that a maximum is reached for $f(x)$, subject to the inequality constraint $g(x) \geq 0$, then *necessary conditions* for p to be a minimum are that a Lagrange multiplier vector $\rho^* \geq \mathbf{0}$ exists, so that the Lagrangian $L(x, \mu) := f(x) + \rho g(x)$ satisfies the *Karush-Kuhn-Tucker conditions* (KKT) or the *saddlepoint condition* (SP):

$$\text{(KKT):} \qquad L_x(p, \rho*) = 0, \rho^* \geq 0, \quad \rho^* g(p) = 0, \; g(p) \geq 0;$$

$$\text{(SP):} \qquad L(\mathbf{p}, \rho) \geq L(p, \mu^*) \geq L(x, \rho^*) \text{ for all x, and all } \rho \geq 0.$$

The conditions often assumed for (KKT) are that f and each component g_i are differentiable functions. If there is also an equality constraint $k(x) = 0$, then a term $\sigma k(x)$ is added to the Lagrangian, and a regularity assumption is required, e.g. that the gradients of the active constraints (those $g_i'(p)$ for which $g_i p) = 0$, together with all the $h_j'(p)$) are linearly independent vectors. The conditions often assumed for (SP) (with k absent) are that f and each g_i are convex functions (which need not be differentiable), together with *Slater's condition,* that $g(z) > 0$ for some feasible point z. Then (SP) is a *sufficient condition* for a minimum (even without assuming convexity). However, (KKT) is not sufficient for a maximum; (KKT) implies a maximum if also f and each g_i are convex functions; this maximum point is unique if f is strictly convex.

If the functions f, g, k also contain a parameter q, then the optimal value of $f(p)$ also depends on q; denote this function by $V(q)$. Under some regularity conditions (see e.g. Fiacco and McCormick, 1968; Craven, 1995), the gradient $V'(\mathbf{0})$ equals the gradient $L_q(p, \rho^*, \sigma^*)$.

For a maximization problem, the inequalities for L in (SP) are reversed, and convexity applies to $-f$ and $-g$.

For the problem with f and g convex functions, and k a linear function, there is associated a *dual problem*:

$$\text{MAX } f(y) + vg(y) + wk(y) \text{ subject to } v \geq 0, f'(y) + vg'(y) + wk'(y) = 0.$$

Assume that the given problem reaches a minimum at $\mathbf{p}$, and that KKT holds with Lagrange multipliers ρ and σ. Then two properties relate the given *primal* problem and the dual problem:

- *Weak Duality* If x satisfies the constraints of the primal problem, and y, v, w satisfy the constraints of the dual problem, then:

$$f(x) \geq f(y) + vg(y) + wh(y);$$

- *Zero Duality Gap (ZDG)* The dual problem reaches a maximum when $(y, v, w) = (\mathbf{p}, \rho, \sigma)$. Thus, the Lagrange multipliers are themselves the solutions of an optimization problem.

These duality properties are well known for convex problems. However, they also hold (see Craven, 1995) when (f, g, k) satisfies a weaker, *invex* property, described in section 2.4. This property holds for some economic and finance models, which are not convex.

2.4. Relaxing the convex assumptions[3]

Convex assumptions are often not satisfied in real-world economic models (see Arrow and Intriligator, 1985). It can happen, however, that a global maximum is known to exist, and there is a unique KKT point; then that KKT point must be the maximum. This happens for a considerable class of economic models – see section 2.5. Otherwise, the necessary KKT conditions imply a maximum under some weaker conditions than convexity. It suffices if f is pseudoconvex, and each g_j is pseudoconcave, or less restrictively if the vector $-(f, g_1, \ldots, g_m)$ is *invex* . (Here k is assumed absent or linear).

A vector function h is *invex* at the point p (see Hanson, 1980; Craven, 1995) if, for some *scale function* η :

$$h(x) - h(p) \geq h'(p)\eta(x, p).$$

(The $\geq$ is replaced by $=$ for a component of k corresponding to an equality constraint.) Note that h is convex at p if $\eta(x, p) = x - p$; but *invex* occurs in other cases as well.

From Hanson (1980), a KKT point p is a minimum, provided that $(f, g_1, \ldots, g_m)$ is invex at p.

If inactive constraints are omitted then (Craven, 2002) this invex property holds exactly when the Lagrangian $L(x, \mu) = f(x) + \mu^T g(x)$ satisfies the *saddlepoint condition* $L(p, \mu) \geq L(p, \mu^*) \geq L(z, \mu^*)$ for all z and all $\mu \geq 0$. If the problem is transformed by $x = \varphi(y)$, where φ is invertible and differentiable, then h is invex exactly when $h \circ \varphi$ is invex (with a different scale function). (Thus, the invex property is invariant to such transformations φ; then name *invex* derives from *invariant convex* .) If a φ can be found such that $h \circ \varphi$ is convex, then it follows that h is invex. (This happens e.g, for the Kendrick-Taylor growth model - see Islam and Craven, 2001a).

For the problem: MAX $f(x)$ subject to $g(x) \geq 0$, where the vector function $(-f, -g)$ is assumed invex, a local minimum is a global minimum; if there are several local minima, they have the same values of $f(x)$. Under the further assumption of *strict invexity* for f at a minimum point p where (KKT) holds (with multiplier μ^*), namely that $f(x) - f(p) > f'(p)\eta(x, p)$ whenever $x \neq p$, a minimum point p is unique. For, with $L(x) = f(x) + \mu^* g(x)$, $L(.)$ is then strictly invex, hence:

$$f(x) - f(p) \geq L(x) - L(p) > L'(p)\eta(x, p) = 0.$$

For optimal control, as in section 2.2, the spaces are infinite dimensional; however KKT conditions (equivalent to Pontryagin, conditions), *quasimax* (see section 2.7), invexity, and related results apply without change. However, *invexity* may be difficult to verify for a control model, because the dynamic equation for $\dot{x}(t)$ is an equality constraint; thus convexity assumptions would require the dynamic equation to be linear.

---

[3] See also sections 4.4 and 8.6.

Invexity can sometimes be established for a control problem by a suitable transformation of the functions, assuming however that constraints on the control function are not active. Consider a model for *optimal growth*, (see Intriligator, 1971; Chiang, 1992), in which $u(t) = c(t)$ is consumption, $x(t) = k(t)$ is capital, $f(x(t), u(t)) = U(c(t))$ where the social welfare function $U(.)$ is positive increasing concave, and:

$$m(x(t), u(t), t) = \varphi(k(t)) - \text{c(t)} - rk(t).$$

Consider now a dynamic equation $\dot{k}(t) = b(t)\theta(k(t)) - c(t)$, where $b(t) > 0$ is given . The transformation $x(t) = \psi(k(t))$ leads to the differential equation:

$$\dot{x}(t) = \psi'(\psi^{-1}(x(t))[b\varphi(\psi^{-1}(x(t)) - c(t))].$$

This assumes that the function $\psi(.)$ is strictly increasing, hence invertible, as well as differentiable. If ψ can be chosen so that $\psi'(.)\varphi(.) = 1$, then the differential equation becomes: $\dot{x}(t) = b(t) - u(t)$, where $u(t) := \psi'(\psi^{-1}(x(t))c(t)$ is a new control function. Then one may ask whether the integrand:

$$-e^{-rt}U(u(t)/\psi'(\psi^{-1}(x(t))$$

happens to be a *convex* function of $(x(t), u(t))$? If it is, then the problem was *invex*, since it could be transformed into a convex problem.

A similar approach was followed by Islam and Craven (2001a) for the Kendrick-Taylor model, defined in section 1.4. Here $\dot{x}(t) = \zeta e^{qt}k(t)^{\beta} - \sigma k(t) - c(t)$ and the integrand is $c(t)^{\tau}$, with $\beta = 0.6$ and $\tau = 0.1$. The transformation $x(t) = (k(t)e^{\sigma t})^{1-\beta}$, followed by $c(t) = x(t)^{\theta}u(t)$ for suitable θ, reduces the dynamic equation to the form:

$$\dot{x}(t) = (1-\beta)\zeta e^{rt} - (1-\beta)e^{\sigma t}u(t).$$

The integrand of the objective function becomes a function of $(x(t), u(t))$, which is concave if its matrix of second derivatives is negative definite. This holds for the given values of β and τ.

The approach of the previous paragraph seems to work a little more generally. However, there are difficulties if $x(t)$ has more than one component; and it must be assumed that any constraints on the control, such the constraint $0 \leq c(t) \leq \varphi(k(t))$ in Chiang (1992), are not active, since these will not transform to anything tractable.

The invex property can sometimes be used to establish existence.

$$\text{MIN } f(x) \text{ subject to } g(x) \leq 0, k(x) = 0,$$

satisfies convex, or invex, assumptions on the functions f, g, h. Without assuming that a minimum is reached, suppose that the necessary KKT conditions at a point $\mathbf{p}$ satisfying the constraints, with some Lagrange multipliers ρ and σ, can be solved for p, ρ, and σ. Define the Lagrangian: $L(x) := f(x) + \rho g(x) + \sigma k(x)$. If x satisfies the constraints, then:

$$f(x) - f(\mathbf{p}) \geq L(x) - L(\mathbf{p} \geq L'(\mathbf{p})\eta(x, \mathbf{p}) = 0.$$

Thus the problem reaches a global minimum at $\mathbf{p}$. This approach is not restricted to a finite number of variables, so it may be applied to optimal control.

It is well known (Mangasarian, 1969) that KKT conditions remain sufficient for an optimum if the convexity assumptions are weakened to $-f$ *pseudoconvex* x and each $-g_i$ *quasiconvex.* The definitions are as follows. The function f is *quasiconcave* at p if $f(x) \geq f(p) => f'(p)(x-p) \geq 0$, *pseudonconcave* at p if $f(x) > f(p) => f'(p)(x-p) > 0$. If f is pseudoconcave, and each g_i is quasiconcave, then a KKT point is a maximum. If the function $n(.)$ is concave and positive, and $d(.)$ is convex and positive, then the ratio $n(.)/d(.)$ is pseudoconcave. Apart from this case (*fractional programming*), quasi- and pseudo-concave are more often assumed than verified. These assumptions can be further weakened as follows. The function h is *quasiinvex* at p if $h(x) \geq h(p) => h'(p)\eta(x,p) \geq 0$, *pseudoinvex* at p if $h(x) > h(p) => h'(p)\eta(x,p) > 0$. As above, the scale function η must be the same for all the functions describing the problem. Then $-f$ *pseudoinvex* and each $-g_i$ *quasiinvex* also make a KKT point a maximum. If $n(.) > 0$ is concave, and $d(.) > 0$ is convex, then a transformed function $-n \circ \varphi(.)/d \circ \varphi(.)$ is pseudoinvex, when φ is differentiable and invertible.

2.5. Can there be several optima?[4]

Many nonlinear programs have local optima that are not global. The evolution in time, in an optimal control model, puts some restrictions on what may happen with such a problem. The following discussion, from Islam and Craven (2004), gives conditions when an optimum is unique, or otherwise. Two classes of control problem often occur:

(a) When f and m are linear in $u(.)$, then *bang-bang control* often occurs, with $u(.)$ switching between its bounds, plus perhaps a singular arc. The optimum is essentially defined by the switching times; and the bounds on $u(.)$ are needed, for an optimum to be reached. Some such problems have several local optima, with different numbers of switching times.

(b) When f and m are nonlinear, and the controls on the constraints are inactive, then the Pontryagin necessary conditions for an optimum requires that:

(i) As well as the dynamic equation, the costate λ (.) satisfies the differential equation :

$$-\dot{\lambda}(t) = e^{-\rho t} f_x(x(t), u(t)) + \lambda(t) m_x(x(t), u(t), t), \quad \lambda(T) = -\Phi'(x(T)),$$

---

4 See also section 5.6.

or, for the Kendrick-Taylor example (see section 2.4):

$$-\dot{\lambda}(t) = e^{-\rho t}U'(c(t)) + \lambda(t)b(t)\theta'(k(t)),\ \ \lambda(T) = -\Phi'(k(T));$$

setting $z(t) = (k(t), \lambda(t))$, the two differential equations combine into one of the form:

$$\dot{z}(t) = \zeta(z(t), u(t), t),\ \ z(0) = z_0;$$

(the λ part of z_0 is a parameter, varied to satisfy the $\lambda(T)$ condition).

(ii) From Pontryagin's principle,

$$e^{-\rho t}f(x(t), u(t)) + \lambda(t)Pm(x(t), u(t), t)$$

is maximized over $u(t)$; or, for the example, $e^{-\rho t}U(c(t)) - \lambda(t)c(t)$ is maximized over $c(t)$ at the optimal $c(t)$; for this (in the absence of constraints on $c(t)$) it is necessary (but not sufficient) that the gradient $e^{-\rho t}U'(c(t)) - \lambda(t) = 0$.

If (I) the gradient equation in (ii) can be solved uniquely, and globally, for $u(t)$, say as

$$u(t) = p(\lambda(t), t),$$

then substitution into the $\dot{z}(t)$ equation gives n

$$\dot{z}(t) = Z(z(t), t) \equiv \zeta(z(t), p(\lambda(t), t), t), z(0) = z_0.$$

Assume additionally (II) that $Z(., t)$ satisfies a Lipschitz condition, uniformly in t. Then the differential equation has a unique solution $z(t)$. In these circumstances, the control problem has a unique optimum. (Note that (I) excludes any jumps in $u(.)$.)

Assumption (I) holds if the utility $U.)$ is concave and strictly increasing. In other cases, there may be several solutions for $u(t)$. In Figures 1 and 2, $U(.)$ is *quasiconcave* , and if $U'(u)$ lies in a certain range there are three solutions for u. However, the optimum control $u(t)$ may still be unique. A simple example, for a pseudoconcave objective (which implies quasiconcave) and a linear differential equation, is given in section 2.8.

In Kurz (1968), a class of optimal control models for economic growth are analysed using the Pontryagin theory. There is a unique optimum if certain concavity conditions are fulfilled. If they are not, then multiple optima may occur. Kurz gives numerical examples of multiple optima when the welfare function $U(.)$ depends on the state $k(.)$ as well as the control $u(.)$.

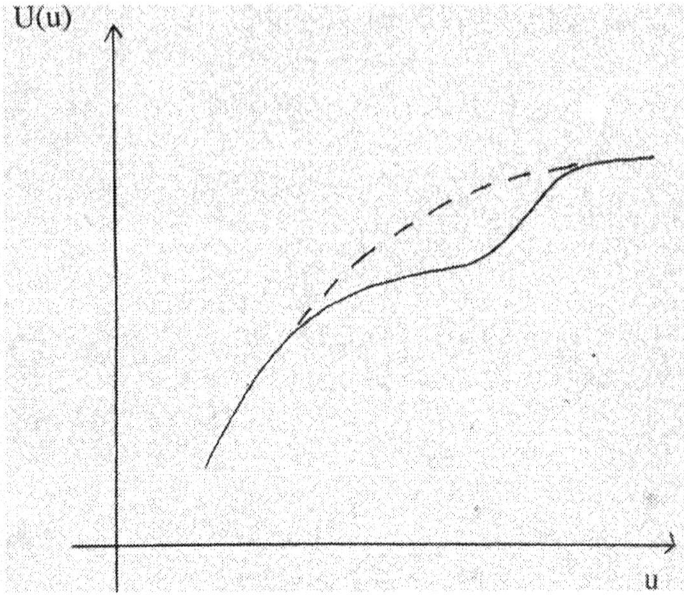

Figure 1

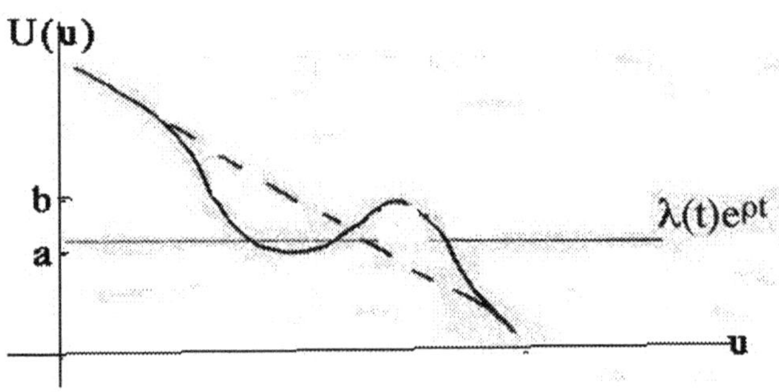

Figure 2

2.6. Jump behaviour with a pseudoconcave objective

An optimal control problem whose objective is pseudoconcave, but not concave, may show jump behaviour in the control, not associated with a boundary of a feasible region. A simple example is proposed in Islam and Craven (2004); it is given here, with numerical results.

Consider the simple example:

$$\text{MAX} \int_0^T U(u(t))dt \text{ subject to:}$$

$$x(0) = x_0 \ , \ \dot{x}(t) = \gamma x(t) - \ \mathrm{u(t)} \ (0 \leq t \leq T), x(T) = x_T \ .$$

The control $u(.)$ may be unconstrained, or there may be a lower bound

$$(\forall t) u(t) \geq u_{lb}.$$

The horizon T is taken as 10, and the growth factor $\gamma = 1$. The quasiconcave (not concave) utility function $U(.)$ is given by:

$$U(u) = u - 0.5u^2 (0 < u < a),$$

$$U(u) = p + (1-a)(u-a) + 0.5d(u-a)^2 (a < u < b),$$

$$U(u) = q + (1+c)(u-b)\_0.5(u^2 - b^2 \ ) \ (u > b) \ ,$$

with parameters chosen to display the jump effect:

$$a = 0.30, b = 0.35, c = 0.10, p = a - 0.5a^2, d = -1 + c/(b-a),$$

$$q = p + (1-a)(b-a) + 0.5d(b-a)^2.$$

This utility function $U(u)$ is constructed from $U(0) = 0$, and $U'(u) = 1-u, 1-a+d(u-a), 1+b-u$ in the three intervals. The increase in slope makes the function U nonconcave. With these numbers, the nonconcavity of U is only just visible on its graph.

In this example $x(t)$ may represent capital, and $u(t)$ may represent consumption, But, when no lower bound is imposed on $u(t)$, negative values for $u(t)$ may be obtained. The model provides substantial capital growth, and it may be advantageous to borrow money to pay for consumption during the early part of the planning period. To model this better, a cost of borrowing :

$$C(u) = \kappa(u - \delta) \text{ when } u < \delta, C(u) = 0 \text{ when } u \geq \delta,$$

may be subtracted from $U(u)$. This replaces an explicit lower bound on $u(t)$. (Here, $\delta = 0.2$ and $\kappa = 1.3$ were considered.) Also, the dynamic equation for $\dot{x}(t)$ is taken here as linear, to enable an analytic solution to the control problem. A more realistic model would replace $\gamma x(t)$ by some increasing concave function of $x(t)$.

From the theory of Section 2.2, the costate $\lambda(t) = -\sigma e^{-\gamma t}$, with some constant σ depending on the terminal condition for $x(T)$; and then $U'(u(t)) = \sigma e^{-\gamma t}$; hence $u(t)$ increases as t increases. Depending on the initial and terminal conditions and the growth factor γ, the range of variation of the optimal

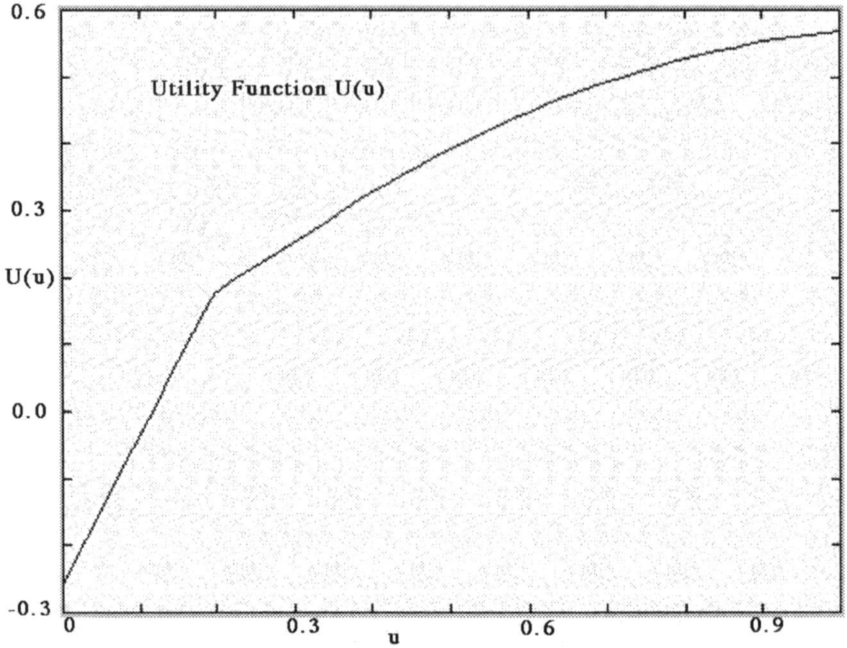

Figure 3 Utility function U(u)

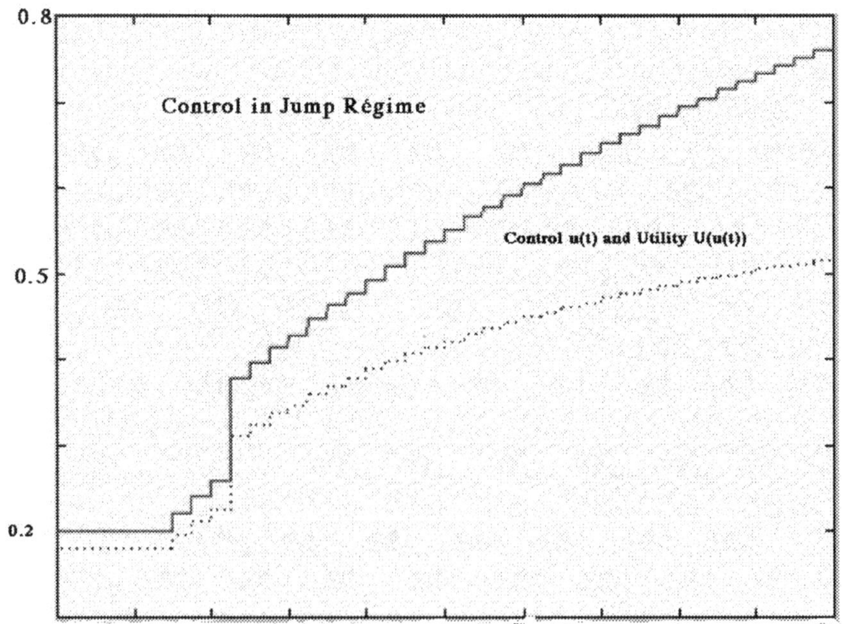

Figure 4 Control in jump regime

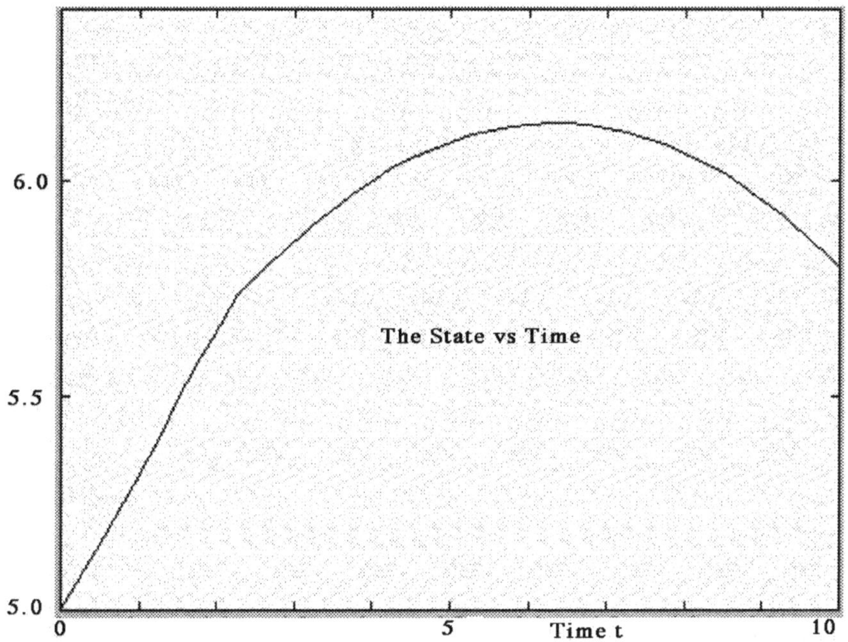

Figure 5 The state

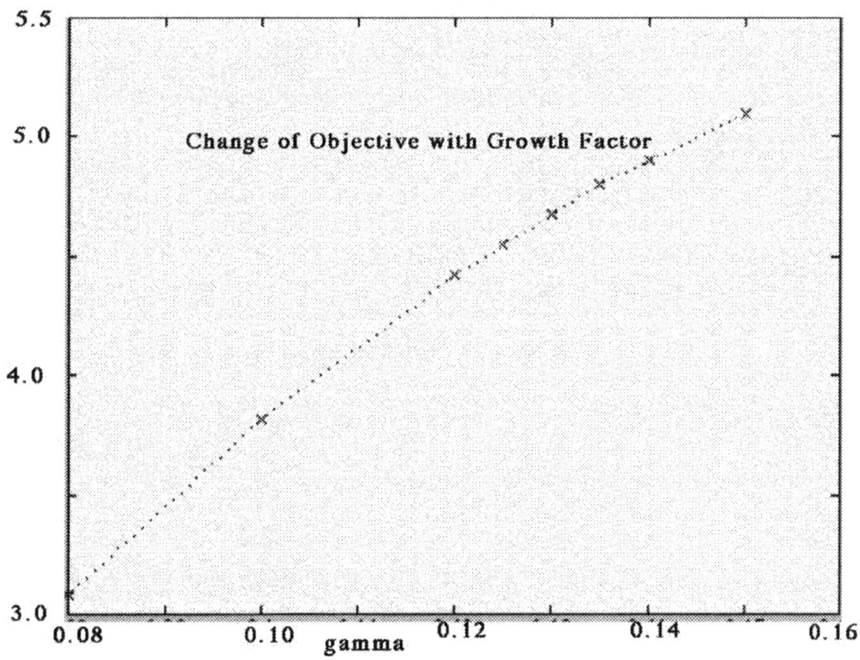

Figure 6 Change of objective with growth factor

$u(t)$ may include the region where there are three possible solutions for $u(t)$. Thus, there are two possible solution regimes, one where this region is not entered, so $u(t)$ varies smoothly, and the second where $u(t)$ jumps from the leftmost part of the $U'(t)$ curve to the rightmost part, at some level of $u(t)$ depending on the parameters mentioned.

Computations with the SCOM optimal control package, described in chapter 3 (also Craven and Islam, 2001) confirm this conclusion. Figure 3 shows the utility function, including the borrowing term. Figure 4 shows the optimal control $u(t)$, and the value of the utility $U(u(t))$, for one set of parameters ($\gamma = 0.10, x0 = 5.0, x_T = 5.80$). In the computation, $uj(t)$ was approximated by a step-function with 40 subintervals; the theoretical solution would be a smooth curve, except for the jump at $t = 2.3$.

Figure 5 shows the state $x(t)$; note the change in slope at $t = 2, 3$, where $u(t)$ jumps.

Figure 6 shows the change in the optimal objective as the growth factor γ changes, other parameters being the same. Note that the control $u(t)$ changes smoothly when $\gamma \leq 0.13$, whereas $u(t)$ has a jump when $\gamma > 0.13$; the graph in Figure 6 changes slope at this level of γ.

2.7. Generalized duality

While *duality* requires convexity, or some weaker property such as *invex*, there are relaxed versions of maximum for which *quasiduality* holds, giving an analog of ZDG, but not weak duality. A point p is a *quasimax* of $f(x)$, subject to $g(x) \leq 0$, if (Craven, 1977)

$$f(x) - f(p) \leq \mathbf{o}(\|x - p\|) \text{ whenever } g(x) \leq 0.$$

Here a function $q(x) = \mathbf{o}(\|x - p\|)$ when $q(x)/\|x - p\| \to 0$ when$\|x - p\| \to 0$. A function f has a *quasimin* at p if $-f$ has a quasimax at p. If f and g are differentiable functions, and a *constraint qualification* holds, then (Craven, 1997) p is a KKT point exactly when p is a quasimax.

Attach to the problem:

$$\text{QUASIMAX}_x f(x) \text{ subject to } g(x) \leq 0$$

the *quasidual* problem:

$$\text{QUASIMIN}_{u,v} f(u) + vg(u) \text{ subject to } v \geq 0,\ \ f'(u) + vg'(u) = 0.$$

Then (Craven, 1977, 1995) there hold the properties:

- *ZDG* If x^* is a quasimax of the primal (quasimax) problem, then there is a quasimin point (u^*, v^*) of the quasidual problem for which $f(x^*) = f(u^*) + v^* gf(u^*)$, thus the objective functions are equal.
- *perturbations* if $V(b) = QUASIMAX f(x)$ subject to $g(x) \geq 0$, with the quasimax for $b = 0$ occurring at $x = x^*$, and if (u^*, v^*) are the

optimal quasidual variables for which ZDG hold with $f(x^*)$, then the *quasi-shadow price* vector $V'(0) = v^*$.

There are several quasimax points with related quasimin points, and they correspond in pairs. These *quasi* properties reduce to the usual ones if $-f$ and $-g$ are *invex* with respect to the same *scale function* $\eta(\cdot)$. However, for an equality constraint $k(x) = 0$, the required invex property has $=$ instead of $\geq$.

Simple examples of *quasimax* and *quasidual* are as follows.

Example 1: (Craven, 1977) The *quasidual* of the problem:

$$\text{QUASIMAX}_x \quad -x + \frac{1}{2}x^2 \text{ subject to } x \geq 0,$$

is:

$$\text{QUASIMIN}_{u,v} \quad -u + \frac{1}{2}u^2 + vu \text{ subject to } -1 + u + v = 0,\ v \geq 0,$$

which reduces to the *quasidual* : $\text{QUASIMIN}_u \quad -\frac{1}{2}u^2$ subject to $u \leq 1$. The primal problem has a quasimax at $x = 0$, with objective value 0, and a quasimin at $x = 1$, with objective value $-\frac{1}{2}$. The quasidual has a quasimin at $u = 0$ with objective value 0, and a quasimin at $u = 1$ with objective value $-\frac{1}{2}$.

Example 2: Here, the quadratic objective may be a utility function for social welfare, not necessarily concave, to be maximized, subject to linear constraints. The quasidual vectors are shadow prices. For given vectors c and s, and matrices A and R, suppose that the problem:

$$\text{QUASIMAX}_{z \in \mathbf{R}^n}\ F(z) := -c^T z + (1/2)z^T Az \text{ subject to } z \geq 0,\ Rz \geq s,$$

reaches a quasimax at a point p where the constraint $z \geq 0$ is inactive, and the constraint $Rz \leq s$ is active. Then KKT conditions give $Ap + R^T\lambda = c,\ Rp = s$, with multiplier $\lambda \geq 0$. Setting $z = p + v,\ F(z) - F(p) = -\lambda^T Rv - (1/2)v^T Av$ with $Rv \geq 0$. So $-\lambda^T Rv \leq 0$, and $F(z) - F(p)$ may take either sign, depending on the matrix A, so p is generally a quasimax, not a maximum. The associated quasidual problem is:

$$\text{QUASIMIN}_{u,v} \quad -c^T u + (1/2)u^T Au - v^T(Tu - s) \text{ subject to}$$

$$c - Au - R^T v = 0,\ v \geq 0.$$

However, the main applicability of *quasimax* is when the primal maximization problem has several local maxima, as is likely to happen when the objective function is far from concave. To each local maximum corresponds a quasimin of the quasidual, with the ZDG property, and quasidual variables, giving shadow prices. Thus the shadow price, for each local maximum, is an optimum point of a quasimin problem.

When there are equality constraints, the *invex* property takes a different form. In general, a vector function F is invex with respect to a convex cone Q at a point p if:

$$(\forall x)\ F(x) - F(p) \in F'(p)\eta(x,p)$$

holds for some *scale function* $\eta(.,.)$. If some component $h(x)$ of $F(x)$ belongs to an equality constraint $h(x) = 0$, then the related part of Q is the zero point $\{0\}$; hence invex requires an equality:

$$(\forall x)\ h(x) - h(p) = h'(p)\eta(x,p).$$

2.8. Multiobjective (Pareto) optimization[5]

When two or more objectives are to be maximized, they usually conflict, so they cannot all be optimal. This is commonly the case with social welfare models. Usually *Pareto* maxima are considered, where moving away from such a point decreases all the objectives. If two objectives F^1 and F^2 should be maximized over a set Γ, define the vector $F(x) := (F^1(x), F^2(x)$. Denote by $\mathbf{W}$ the set of points $w \in \mathbf{R}^2$ for which $w_1 > 0$ and $w > 0$ do not both hold; $\mathbf{W}$ is shown in the right diagram. Then p is a *Pareto maximum* if, for each $x \in \Gamma$, F(x) $-F(p) \in \mathbf{V}$, where $\mathbf{V}$ is the set indicated in the left diagram, namely the interior of $\mathbf{W}$, together with 0. The point p is a *weak maximum* if, for each $x \in \Gamma$, F(x) $-F(p)$ does not lie in $\mathbf{W}$. The weak maximum points include all the Pareto maximum points, and some additional boundary points, excluded from $\mathbf{V}$. For $r > 2$ objectives, $\mathbf{R}^r$ replaces $\mathbf{R}^2$.

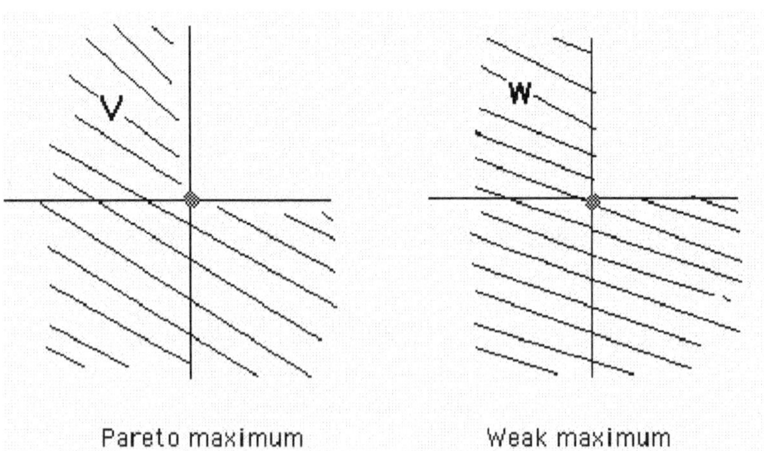

More generally, consider the "*maximization*" of a vector objective function $F(.)$ subject to the constraint $g(.) \geq 0$, relative to an order cone Q in the space

[5] See sections 6.4 and 8.4.

into which F maps. Denote by $\mathbf{U}$ the complement of the interior of Q. A point p is a *weak quasimax* of $F(.)$, subject to $g(.) \geq 0$, if for some function $q(x) = o(||x - p||)$, there holds:

$$F(x) - F(p) - q(x) \in \mathbf{U} \text{ whenever } g(x) \geq 0.$$

(Note that *weak maximum* is the special case $q(.) = 0$.) Replacing $\mathbf{U}$ by the union of 0 with the interior of $\mathbf{U}$ gives the slightly more restrictive *Pareto maximum;* a *weak maximum* may include a few additional boundary points. A *weak quasimin* of a vector function H is a weak quasimax of -H . If F has r components, then usually Q is taken as the orthant $\mathbf{R}^r_+$. The following diagram (see Craven, 1981) illustrates the definition, for the case when F has two components, so that the order cone Q is a sector, but not always the non-negative orthant $\mathbf{R}^2_+$. Such an ordering is appropriate when the consumers are not isolated, as is discussed in section 8.2.

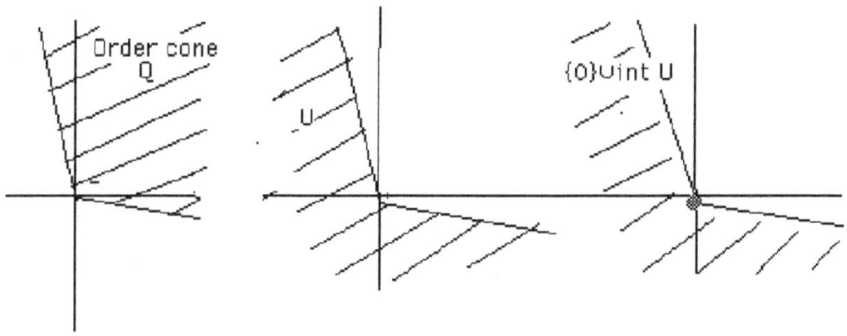

From Craven (1995), a (maximizing) *weak Karush-Kuhn-Tucker point (WKKT)* p satisfies:

$$\tau F'(p) + \lambda g'(p) = 0, \lambda g(p) = 0, \lambda \geq 0\ \tau \geq 0, \tau e = 1.$$

Here τ is a row vector, e is a column vector of ones. The statement $\tau \geq 0$ means, more precisely, that τ lies in the dual cone Q^* of Q, defined by $q^*(q) \geq 0$ whenever $q^* \in Q^*$ and $q \in Q$, If $Q = \mathbf{R}^r_+$ then each component of τ is non-negative. There are many WKKT points, corresponding to different values of τ.

Analogously to section 2.7 (and see Craven, 1990), the *weak quasidual* to the (primal) problem:

$$\text{WEAK QUASIMAX}_x F(x) \text{ subject to } g(x) \geq 0$$

has the form:

$$\text{WEAK QUASIMIN}_{u,V} F(u) + Vg(u) \text{ subject to}$$

$$V(\mathbf{R}^m_+) \subset Q, (F'(u) + Vg'(u))(X) \subset \mathbf{U},$$

where the space $X = \mathbf{R}^n$ is the domain of F and g, and V is a $r \times m$ matrix variable. The *linearized problem* about the point p:

$$\text{QUASIMAX } F'(p)(x-p) \text{ subject to } g(p) + g'(p)(x-p) \geq 0$$

If a constraint qualification holds, and without any convex or invex assumption, there hold (Craven, 1989, 1990):

- p is a WKKT point exactly when p is a weak quasimin.
- *ZDG* If x^* is a weak quasimax point of the primal, with multipliers τ, λ, then there there is a weak quasimax point $(u, V) = (x^*, V^*)$ of the weak quasidual problem, with $\tau V^* = \lambda$, and equal optimal objectives:

$$F(x^*) = F(u^*) + V^* g(u^*).$$

Proof Let x and u, V be feasible for the respective problems. Then (using quasidual constraints):

$$[F(u) + Vg(u)] - [F(x^*) + V^*g(x^*)) \in Q + [F(u) + V^*g(u)]$$

$$-[F(x^*) + V^*g(x^*) + (V - V^*)(g(u) - g(x^*)\}$$

$$= Q + (F + Vg)'(x^*)(u - x^*) + o(\|u - x^*\|) + (o(\|u - x^*\|) + o(\|V - V^*\|)),$$

$$\subset \mathbf{U} + o(\|u - x^*\| + \|V - V^*\|).$$

- *Linearization* The linearized problem about p reaches a weak maximum at p.
- *Shadow prices* If x^* is a weak quasimax of the primal, then x^* is also a quasimax of $\tau^* F(x)$; so the shadow prices for $\tau^* F$ are optimal for:

$$\text{QUASIMAX } F'(p)(x-p) \text{ subject to } g(p) + g'(p)(x-p) \geq 0$$

- *Multilinear problem* (Bolinteanu and Craven, 1992) If F is linear, and g is linear (plus a constant), then p is stable to small perturbations, and there is a shadow price for F for perturbations that do not change the list of active constraints.

Example 3 - Vector quasimax and quasidual

$$\text{WEAK QUASIMAX}_{z \in \mathbf{R}^n} \{F_i(z)\}_{i=1}^r := \{-c_i^T z + (1/2) z^T A_i iz\}_{i=1}^r,$$

$$\text{subject to } Rz \geq s,$$

for given vectors c_i, d, s and matrices A_i, R; here $\{F_i(z)\}_{i=1}^r$ specifies a vector objective by its components. For each multiplier $\tau \geq 0$, with $e^T \tau = 1$, where e is a vector of ones, there corresponds a quasimax $z = z(\tau)$, with:

$$A^T z(\tau) + R_0^T \lambda(\tau) = c, \;\; R_0 z(\tau) = s_0,$$

in which $R_o z \geq s_0$ describes the constraints active at $z(\tau)$, $c = \sum \tau_i c_i$, $A = \sum \tau_i A_i$, As in example 2, this quasimax becomes a maximum if the matrix A is restricted, e.g. to be negative definite in feasible directions. This problem may describe e.g. conflicting objectives of output and envoronmental quality for sustainable development, and requiring nonlinear functions to describe them. The vector quasidual is then:

$$\text{WEAK QUASIMAX}_{u,V} \; \{-c_i + u^T A_i\}_{i=1}^r + VR)(\mathbf{R}^n) \subset \mathbf{U}, \; V \geq 0,$$

where $\mathbf{U} = \mathbf{R}^r \backslash$ int $\mathbf{R}^r_+$; each component of V is ≥ 0.

While a *quasimax* is not generally a maximum, the *quasi* properties reduce to the standard properties if an additional assumption, such as *convex* or *invex*, is made. Consider the problem of weak maximization of $F(x) <$ *with* respect to the order cone Q, subject to constraints $g(x) \geq 0$ and $k(x) = 0$. In weak KKT, a term $\sigma k'(p)$ is added to $\lambda g'(p)$, where the multiplier σ is not restricted in sign. The *invex* property stated in section 3.2 (see Hanson, 1980; Craven, 1981; Craven, 1995) now takes the form:

$$-F(x) + F(p) + F'(p)\eta(x,p) \in Q,$$

$$g(x) - g(p) \leq g'(p)\eta(x,p),$$
$$k(x) - k(p) = k'(p)\eta(x,p),$$

for the same *scale function* $\eta(x,p)$ in each case. (The inequalities derive from a *cone-invex* property $H(x) - H(p) - H'(p)\eta(x,p) \in Q \times \mathbf{R}^m_+ \times \{0\}$ for a vector function $H = -(F, g, k)$ (see Craven 1995).

In particular, if all components of F and g are concave, and all components of k are affine (constant plus linear), then invexity holds with η (x, p) $= x - p$, and the necessary conditions *weak KKT*, or equivalently *weak quasimax,* become also sufficient for a maximum. But these assumptions are often not fulfilled in applications. Some conditions when invexity holds are given in Craven (1995, 2000). A transformation $x = \varphi$ (y) of the domain, where φ is invertible, with both φ and φ^{-1} differentiable, preserves the *invex* property, although the scale function is changed. The name *invex* derives (Craven, 1981) from *invariant convex* ; thus if a function $\mathbf{H}$ is convex, then the transformed function $\mathbf{H}_\circ \varphi$ is *invex* (though not every invex function is of this form). However, for an equality constraint $k(\ \mathbf{x}\ ) = 0, k(.)$ is invex at p if $k(x) - k(p) = k'(p)\eta$ (x, p), For KKT to be sufficient for an optimum, this invex property is only required for those x where $k(x) = 0$. So the requirement reduces to the *reduced invex (rinvex)* property: $k(x) = 0 = k'(p)\eta(x,p)$.

A multiobjective analog of *saddlepoint* was given in Craven (1990), and the results are summarized here. The point p is a *weak saddlepoint* (WSP) (see Craven, 1990) if:

$$\tau^T F(p) + v^T g(p) \geq \tau^T F(p) + \lambda^T g(p) \geq \tau^T F(x) + \lambda^T g(x)$$

holds for all $\mathbf{x}$, and all $v \geq 0$. The multiplier τ (where $0 \neq \tau \geq 0$) depends on p. The point p is a *weak quasisaddlepoint* if (given τ):

$$(\forall x, \forall \mu \geq 0)\tau F(p) + \mu g(p) \geq \tau F(p) + \lambda g(p) \geq \tau F(x) + \lambda g(x) + \mathbf{o}(\|x - p\|).$$

The following relations hold (Craven, 1995) for the vector function F on E and $p \in E$:

p is a Pareto maximum $\Rightarrow p$ is a weak maximum

$\Rightarrow p$ is a weak quasimax

$\Leftrightarrow p$ is a weak KKT point (assuming regularity of the constraint)

$\Leftrightarrow p$ is a weak quasisaddlepoint .

Weak maximum (and *weak quasimax*) points may include some boundary points which lack a desired stability property. The statement that $0 = \tau = (\tau_1, \ldots, \tau_r) \geq 0$ means that all components $\tau_i \geq 0$, and they are not all zero. If $\tau > 0$, namely if all $\tau_i > 0$, then (Craven, 1990): weak KKT & $\tau > 0 \Leftrightarrow$ *locally proper weak quasimax* , where *locally proper* means that ratios of deviations of the different $\mathbf{F}_i$ from $\mathbf{F}_i p)$ are bounded (inequalities of Geoffrion, 1968) when x is near p. These ratios represent *tradeoffs* between the objective components. They are approximated by the corresponding ratios of the τ_i. Thus a *proper* quasimax occurs exactly when all the multipliers τ_i for the objectives are strictly positive. This property does not require the Lagrange multipliers to be stated. As motivation, note that the *stationary points* of a real function f are the point p where the gradient $f'(p) = 0$; and maximum points are sought among the stationary points, If there are constraints, then *constrained maximum points* are among the KKT points instead of stationary points, A quasimax point p is a maximum point of the linear program, obtained by linearizing the given problem about the point p.

2.9. Multiobjective optimal control

For an optimal control problem with a single objective, the KKT conditions are essentially equivalent to an adjoint differential equation for a costate function, together with Pontryagin's principle. This extends to multiobjective control problems.

Consider an optimal control problem, to find Pareto maximum points of a vector objective $F(x, u)$, subject to a dynamic equation that determines the state x in terms of the control u, and other constraints. By substituting for x in terms of u, the problem takes the form:

$$\text{PMAX } \mathbf{J}(u) \text{ subject to } u \in \mathbf{U},$$

where PMAX denotes Pareto maximum, and the feasible set $\mathbf{U}$ is a closed bounded subset of some vector space $\mathbf{V}$. Suppose that $u \in \mathbf{U} \Leftrightarrow \mathbf{G}(u) \leq 0$, for some function $\mathbf{G}$. If a Pareto maximum point $\hat{u}$ is reached, then (under restrictions usually satisfied), multipliers τ and ρ exist such that Karush-Kuhn-Tucker (KKT) conditions hold:

$$\tau\mathbf{J}'(\hat{u}) + \rho\mathbf{G}'(\hat{u}) = 0, 0 \neq \tau \geq 0, \rho \geq 0, \rho\mathbf{G}(\hat{u}) = 0.$$

If $-\mathbf{J}$ and $\mathbf{G}$ satisfy some relaxed version of convexity, such as *invexity* (see Craven, 1995), then the KKT conditions, in turn, imply a Pareto maximum, and also that $\tau\mathbf{J}(u)$ is maximized, subject to $\mathbf{G}(u) \leq 0$, at $\hat{u}$. Then each τ corresponds to one or more Pareto maximum points, and each Pareto maximum point corresponds to one or more multipliers τ. If minimum points of $\tau\mathbf{J}(.)$, subject to $\mathbf{G}(u) \leq 0$, exist for each $\tau \geq 0$, then the existence of Pareto maximum points is established.

For an optimal control problem in continuous time, the vector space $\mathbf{V}$ is infinite dimensional, so compactness properties of $\mathbf{U}$ are not generally available, so a Pareto maximum of a continuous vector function is not always reached. For each given τ, assume that $J(u) := \tau\mathbf{J}(u)$ has a finite lower bound over $\mathbf{U}$. If $J(u)$ has the form $\int_0^T f(x(t), u(t), t)dt$, with a *coercivity* assumption on the integrand f, that:

$$(\forall x, u, t) f(x, u, t) \geq g(u) \text{ , where } g(u)/|u| \to \infty \text{ as } |u| \to \infty,$$

together with convexity of a set: $\cup_{u \in U}\{$ (z, y) : $z = m(x, u, t)y \geq f(x, u, t)$, where U is the closed set of values allowed to $u(t)$, then a maximum for $J(u)$ exists (see Fleming and Rishel, 1975, Theorem 4.1). But these assumptions are often not satisfied for an economic growth model.

Suppose, however, that $u = p$ is a solution of the KKT necessary conditions for a maximum of $J(u)$ (or equivalently of the Pontryagin conditions – see section 2.2). Additional assumptions are needed, to ensure that p is a maximum. Suppose that $S_1 \subset S_2 \subset \ldots \subset S_n \subset \ldots$ is an increasing sequence of finite-dimensional subspaces of $\mathbf{V}$, and that p is a limit point of their union. Suppose also that each $S_n \cap \mathbf{U}$ is bounded (in a finite dimension); then $J(u)$ reaches a maximum, say at $u = p_n$, over $S_n \cap \mathbf{U}$. Then $J(p_n) \to J(p)$, and p is a maximum point for J. (See Craven (1999b) for an example for a control model where bang-bang control is optimal.)

Otherwise, if $u = p$ satisfies the Pontryagin conditions, and if the functions of the problem satisfy *invex* conditions (see Craven 1995), then p is a maximum. In particular, it suffices if the integrand f of the objective is concave, and the dynamic differential equation for the state is linear. Less restrictively, a function φ is *invex* at p if :

$$\varphi(u) - \varphi(p) \geq \varphi'(p)\eta(u, p),$$

where the *scale function* η must be the same for the several functions (-f and constraint functions) describing the problem. But this is often not the case for a given dynamic equation. However, the control problem has a *local* maximum at p if $-f$ is invex, under lesser restrictions on the differential equation (see section 8.6.)

The maximum point p for $J(u)$ is unique if the Pontryagin conditions have a unique solution. This may happen, in particular (see Islam and Craven, 2001b) when there are no control constraints, so that the Pontryagin maximum condition reduces to a gradient equation, $f_u + \lambda m_u = 0$. If this equation defines u as a unique function of λ, and some Lipschitz conditions hold for f, m, m_x and m_u , then the differential equations for $\dot{x}$ and $\dot{\lambda}$ (substituting for u) have unique solutions.

For a multiobjective problem, the maximum point cannot be unique, but the set of Pareto maximum points is unique, if one of the above criteria applies.

2.10. Multiobjective Pontryagin conditions[6]

Consider an optimal control problem, to find a Pareto (or weak) maximum for two objective functions over a time interval [0, T] :

$$\text{PMAX } \{ \int_0^T f^1(x(t), u(t), t)dt + \Phi^1(x(T)),$$

$$\int_0^T f^2(x(t), u(t), t)dt + \Phi^2(x(T))\}$$

subject to:

$$x(0) = x_0, \dot{x}(t) = m(x(t), u(t), t) \ (0 \ \leq t \leq T),$$

$$a(t) \leq u(t) \leq b(t) \ (0 \ \leq t \leq T).$$

Here the state function $x(.)$ and control function $u(.)$ can be scalar or vector valued; PMAX denotes Pareto maximum; there may be endpoint contributions Φ_1, Φ_2. Note that a Pareto minimum of $\{F^1, F^2\}$ is a Pareto maximum of $\{-F^1, -F^2\}$. There can be more than two objectives.

The relevance of such models to economic growth, decentralization, social choice, and sustainable growth is discussed in chapters 5, 6 and 8. In particular, components of $u(.)$ may represent consumption, and components of $x(.)$ may represent capital stock, with a dynamic differential equation describing economic growth. Questions of existence and uniqueness are discussed in sections 2.3 and 2.5.

The above optimal control model can be described by a *Hamiltonian* :

$$h(x(t), u(t), t; \tau, \lambda(t)) :=$$

---

[6] See section 8.4.

$$\sum_i \tau_i[f^i(x(t), u(t), t) + \delta(t - T)\Phi^i(x(T))] + \lambda(t)m(x(t), u(t), t)$$

Here $\lambda(.)$ is the *costate function* ; τ_i are nonnegative multipliers; and the end-point terms $\Phi^i(x(T))$ have been replaced by terms $\delta(t - T)\Phi^i(x(T))$ added to the integrands; $\delta(.)$ is Dirac's delta-function.

The analysis in Craven (1999a) uses a *Lagrangian* $L(.) = h(.) + \dot{\lambda}(t)x(t)$.

The optimum is then described by the necessary Karush-Kuhn-Tucker (KKT) conditions. From this is deduced, under some regularity assumptions, a *(weak) adjoint differential equation*:

$$\dot{\lambda}(t) = \tau^T f_x(x(t), u(t), t) + \lambda(t)m_x(x(t), u(t), t), \lambda(T) = \tau^T \Phi'(x(T)),$$

(where f_x denotes $(\partial/\partial x)f$, etc.), together with the Pontryagin principle:

$$h(x(t), ., t; \tau, \lambda(t)) \rightarrow \text{ MAX over } [a(t), b(t)] \text{ at the optimal } u(t).$$

There are many Pareto (or weak) optimal points, corresponding to various values of τ, where $0 \neq \tau \geq 0$. If the constraints on the controls are inactive, then MAX h implies:

$$h_u(x(t), u(t), t; \tau, \lambda(t)) = 0$$

at the optimal $x(t), u(t), \lambda(t)$. (Note that this does not imply MAX h.)

The costate function $\lambda(t)$ represents a *shadow price* for the differential equation. Thus, if the dynamic differential equation is perturbed to:

$$x(0) = x_0, \dot{x}(t) = m(x(t), u(t), t) - \beta(t),$$

where $\beta(.)$ is a continuous function of small norm, then the change in the optimal value of $\tau f(.)$ is approximated by $\int_0^T \lambda(t)\beta(t)dt$.

An alternative description uses a *vector Hamiltonian* :

$$\mathbf{H}(x(t), u(t), t; \Lambda(t)) :=$$

$$\mathbf{f}(x(t), u(t), t) + \delta(t - T)\Phi(x(T)) + \Lambda(t)m(x(t), u(t), t),$$

in which **f** is the vector of f^i, Φ is the vector of φ^i , and $\Lambda(t)$ is a matrix valued function instead of the vector function $\lambda(.)$. The *vector adjoint equation* is then:

$$\dot{\Lambda}(t) = f_x(x(t), u(t), t) + \Lambda(t)m_x(x(t), u(t), t), \Lambda(T) = \Phi'(x(T)).$$

Corresponding to Pontryagin's principle, there is a Pareto (or weak) optimum of the vector Hamiltonian, thus:

$$\mathbf{H}(x(t), ., t; \Lambda(t)) \rightarrow \text{ PMAX over } [a(t), b(t)] \text{ at the optimal } u(t).$$

The *maximum* in Pontryagin's principle becomes here a *Pareto maximum.* If the constraints on the controls are inactive, then optimizing the vector Hamiltonion implies (for two objectives) that:

$$(\exists\tau \geq 0, \tau \neq 0)\tau\mathbf{H}_u(x(t), u(t), t; \Lambda(t)) = 0,$$

at the optimal $x(t), u(t), \Lambda(t)$. (Note that this does not imply PMAX.)

If multipliers τ and λ are known, then $\Lambda(t)$ may be constructed from $\tau^T\Lambda(t) = \lambda(t)$.

Chapter 3
Computing Optimal Control : The SCOM package

3.1. Formulation and Computational Approach

This chapter discusses computational approaches to optimal control in economics and finance, in particular for modelling social choice and public decision making. For the mathematical background, refer to sections 2.1 and 2.2. A computing package called SCOM has been developed by the present authors, using MATLAB (see sections 3.3 and 3.4, and Craven and Islam, 2001). Some applications are given in sections 3.6 and 3.7.

Consider an optimal control problem, of the form:

$$\mathrm{MIN}_{x(.),u(.)}\ F^0(x,u) := \int_0^1 f(x(t),u(t),t)dt + \Phi(x(1))$$

subject to $x(0) = a,\ \dot{x}(r) = m(x(t),u(t),t),\ q(t) \leq u(t) \leq r(t)\ \ 0 \leq t \leq 1).$

Here the state function $x(.)$, assumed piecewise smooth, and the control function $u(.)$, assumed piecewise continuous, are, in general, vector-valued; the inequalities are pointwise. A substantial class of optimal control problems can (see Teo et al., 1991; Craven, 1998) be put into this form; and, in many cases, the control function can be sufficiently approximated by a step-function. A terminal constraint $\sigma(x(1)) = b$ can be handled by replacing it by a penalty term added to $F^0(x,u)$; thus the objective becomes:

$$F(x,u) := F^0(x,u) + \tfrac{1}{2}\mu\|\sigma(x(1)) - b^*\|^2,$$

where μ is a positive parameter, and b^* approximates to b.

In the augmented Lagrangian algorithm (see e.g. Craven, 1995), constraints are thus replaced by penalty terms; μ is finite, and typically need not be large; here $b^* = b + \theta/\mu$, where θ is a Lagrange multiplier. If there are few constraints (or one, as here), the problem may be considered as one of parametric optimization, varying b^*, without computing the multipliers.

The differential equation, with initial condition, determines $x(.)$ from $u(.)$; denote this by $x(t) = Q(u)(t)$; then the objective becomes:

$$J(u) = F^0(Q(u),u) + \tfrac{1}{2}\mu\|\sigma(Q(u)(1)) - b^*\|^2,$$

For the formula to compute the gradient $J'(u)$ of the objective, refer to section 2.1.

A diversity of algorithms have been proposed for computing optimal control. Algorithms for optimal control (Mufti, 1970, Teo et al., 1991) have been variously based on: (i) dynamic programming; (ii) solving first-order necessary conditions of the Pontryagin theory; (iii) applying some approximation methods (steady state solution, numerical methods based on approximation and perturbation, and method of simulation); (iv) approximating the control; and (v) applying mathematical programming algorithms to a discretized version of the control problem.

Many of these methods (especially methods (i) to (iii)) are based on the first order maximization conditions of a control problem, which Schwartz (1996) has called indirect methods. About the suitability of these methods for computing optimal control, Schwartz (1996, p. 1) has made the following statement:

The main drawback to indirect methods is their extreme lack of robustness: the iterations of an indirect method must start close, sometimes very close, to a local solution in order to solve the two-point boundary value subproblems. Additionally, since first order optimality conditions are satisfied by maximisers and saddle points as well as minimizers there is no reason, in general, to expect solutions obtained by indirect methods to be minimizers.

Against these indirect methods, he has advocated the superiority of the direct methods, the principles of which, according to him are as follows (p. 1):

Direct methods obtain solutions through the direct minimization of the objective function (subject to constraints) of the optimal control problem. In this way the optimal control problem is treated as an infinite dimensional mathematical programming problem.

The superiority of this direct method or the control discretisation method has also been advocated and implemented increasingly by others in recent years (Craven, 1998 ; Teo et al., 1991). The direct method considers an optimal control problem as an mathematical programming problem in infinite dimensions (see e.g. Tabak and Kuo, 1971), and then discretizes to permit computation.

The discretisation method has also several advantages: relatively large-scale and complicated optimal control problems can be solved by this approach and the method is relatively accurate and efficient compared to the indirect methods.

For the above reasons, we have adopted the discretisation (step function and spline) approach in this book. The two computer programs, SCOM and RIOTS_95, used in this chapter for numerical computation are also based on this approach.

The problem must be put into the above standard form, for convenience scaling the time interval to $[0, 1]$.. As described in Craven, de Haas, and Wettenhall (1998), a nonlinear scaling of time is sometimes appropriate, to get good accuracy when the horizon T is large, without excessive computation.

For SCOM, the interval $[0, 1]$ is now divided into N equal subintervals, and $u(t)$ the successive subintervals. Thus, the discretization is achieved by restricting $u(.)$ to such a step-function, and the optimization is over the values

$u_1, u_2, \ldots, u_N$. Then, from the differential equation, $x(.)$ is a polygonal function, determined by its values $x_0, x_1, \ldots, x_N$ at the gridpoints $t = 0, 1/N, 2/N, \ldots, 1$. Since, because of the jumps at gridpoints, the right hand side of the dynamic differential equation is now not a smooth function of t the differential equation solver chosen must be suitable for such functions. From the standard theory, the gradient $J'(u)$ is given by:

$$J'(u)z = \int_0^1 (f + \lambda(t)m)_u(x(t), u(t), t)z(t)dt,$$

where the costate function $\lambda(.)$ satisfies the adjoint differential equation:

$$-\dot{\lambda}(t) = (f + \lambda(t)m)_x(x(t), u(t), t), |\lambda(1) = \mu(x(1) - b^*).$$

(Here suffixes $_x$ and $_u$ denote gradients with respect to $x(.)$ and $u(.)$.

3.2. Computational requirements

The computer package SCOM is compared with another computer package, RIOTS_95, which is also based on MATLAB. The algorithmic steps for these two packages may be summarized as follows:

- SCOM: approximation based on a step function, a differential equation solution based on Runge-Kutta method and optimisation based on a onjugate gradient search method.
- RIOTS: package uses various spline approximations to do this, and solves the differential equations by Runge Kutta methods and the optimisation problem by projected descent.

In solving the adjoint differential equation , interpolation is required, to estimate values of $x(t)$ at times t between gridpoints. (In *SCOM*, linear interpolation was used.) Similarly,

$$\partial J(u)/\partial u_j = \int_{j/N}^{(j+1)/N} (f + \lambda(t)m)_u(x(t), u(t), t)$$

requires either interpolation of $x(t)$ and $\lambda(t)$, or an Euler approximation of the integral. (SCOM uses the latter; it may be too approximate for some problems.) Of course, the number N of subintervals can be increased, to get better precision.

The differential equation solver must handle discontinuities at gridpoints. Many standard solvers do not. For example, MATLAB 5.2 includes six ordinary differential equation (ODE) solvers, of which only one – designated for stiff differential equations – is useful for solving $\dot{x}(t) = u(t)$ when $u(.)$ is a step-function. A better approach is to modify slightly the well-known fourth order Runge-Kutta method. If t is in the subinterval $[j/N, (j+1)/N]$, then $u(t)$ must take the appropriate value u_j, and not (for example) u_{j+1} when $t = (j+1)/N$. This is easily achieved by recording j as well as t. With this approach, there is no need to further divide the subintervals, in order to integrate. If more precision is required, the number N of subintervals may be increased. Note that the given differential equation is solved forwards (starting at $t = 0$), and the adjoint differential equation is solved backwards (starting at $t = 1$). In solving the adjoint equation, $x(t)$ must be interpolated between gridpoints; in computing $J'(u)$, both $x(t)$ and $\lambda(t)$ must be interpolated; in both cases, linear interpolation was used. More precision can be got by increasing N.

Once the differential equation for $x(.)$ is solved, the objective $J(u)$ becomes a function $\tilde{J}(u_1, u_2, \ldots, u_N)$ of N variables. The optimization may be done, either:

- by computing objective values, but not gradients; or
- by also computing gradients, using the adjoint differential equation.

In either case, the computing package is required to optimize the objective, subject to simple bounds. (Bounds could also be handled using penalty terms.)

When as in some economic models (see e.g. Kendrick and Taylor, 1971), fractional powers of the functions occur, e.g. with the term $x(t)^\beta$ included in $m(x(t), u(t), t)$ with $0(\beta < 1$, then the differential equation (2) will bring $x(t) < 0$ for some choices of $u(.)$, causing the solver to crash. Of course, only positive values of $x(t)$ have meaning in such models. Even without explicit constraints on $u(.)$, the requirement $x(.) > 0$ forms an implicit constraint, which does not have a simple form. In such a case, it may be better to use finite-difference approximations to gradients, since they may be valid, as useful approximations, over a wider domain.

Every algorithm for numerical computation of such an optimal control model requires some approximation of the control function $u(t)$ by a vector on some vector space of finite dimension (n say). For comparison, the RIOTS_95 package uses various spline approximations to do this (Schwartz, 1996), and solves the differential equations by projected descent methods. A simpler approach, followed by the MISER3 package for optimal control (Jennings et al., 1998), approximates the control function by a step-function, then solves the differential equations by sequential quadratic programming. Conjugate gradient methods may also be used. e.g. with OCIM (Craven, de Haas and Wettenhall, 1998). Different implementations may behave differently, in particular if the function is only defined on a restricted domain, since an optimization method may want to search outside that domain. While a step-function is apparently a crude approximation, it has been shown in various instances (see Teo et al., 1991) to produce accurate results. The reason is that integrating the dynamic equation $\dot{x}(t) = \ldots$ to obtain $x(t)$ is a smoothing operation, which attenuates high-frequency oscillations. It is pointed out in Craven (1995) that, if this attenuation is sufficiently rapid, the result of step-function approximations converges to the exact optimum as $N \to \infty$.. Some assumption of this qualitative kind is in any case necessary, in order to ensure that the chosen finite dimensional approximations shall allow a good approximation to the exact optimum.

Computer programs which can be used for a wide variety of continuous control problems have become available only recently (Amman, Kendrick, and Rust, 1996; Cesar, 1994; Tapiero, 1998; Teo et al., 1991). There are some packages which are well known to economists such as DUAL, MATHEMATICA, etc.; see references in Amman, Kendrick and Rust (1996). Craven (1995) covers the theory and algorithm for optimal control with a list of possible computer programs including the OCIM program. Cesar (1994) also has references to some other computer programs. A recent survey of optimal control packages is also given in Tapiero (1998). RIOTS is developed by Schwartz, Polak and Chen (1997).

The various optimisation algorithms that some of the optimal control programs encode are as follows:

• sequential quadratic programming (MISER, Jennings et al. 1991), MATLABś optimiser *constr* (MATLAB, 1997), and SCOM);

• projected descent methods (RIOTS_95, Schwartz 1996).

They can behave differently, in particular if the function is only defined on a restricted domain, since an optimisation method may want to search outside that domain. For example, the RIOTS_95 package uses various spline approximations to do this, and solves the differential equations by projected descent methods. A simpler approach, followed by the MISER3 package for optimal control, approximates the control function by a step-function, then solves the differential equations by sequential quadratic programming. Conjugate gradient methods may also be used (as in OCIM). Different implementations may behave differently, in particular if the function is only defined over a restricted domain, since an optimisation method may want to search outside that domain. While a step-function is apparently a crude approximation, it has been shown in various instances (e.g. Craven 1995; Teo et al. 1991) to produce accurate results. The reason is that integrating the dynamic equation ($\dot{x}(t) = \ldots$) to obtain $x(t)$ is a smoothing operation, which attenuates high-frequency oscillations. It is pointed out in Craven (1995) that if this attenuation is sufficiently rapid, the result of step-function approximations converges to the exact optimum as $N \to \infty$. Some assumption of this qualitative kind is in any case necessary, in order to ensure that the chosen finite dimensional approximations shall allow a good approximation to the exact optimum.

From the alternatives discussed above, RIOTS and SCOM were chosen, because:

• these two programs can solve, easily and conveniently, a large class of optimal control problems with various types of objective functionals, differential equations, constraints on the state and control functions, and terminal conditions (such as bang bang, minmax, and optimal time control problems; see for the theory Sengupta and Fanchon, 1969; Teo, Goh and Wong, 1991) ,

• they are based on a widely used mathematical package MATLAB, and

• RIOTS is commercially available for research in this area.

3.3. Using the SCOM package

The details of SCOM are given in the next section 3.4. It was written to use MATLAB version 5.2, and implemented on a Macintosh computer. Since MATLAB is designed for matrix computations, vector and matrix operations require very little programming. MATLAB's Optimization Toolbox includes a constrained optimizer, *constr* , based on a sequential quadratic programming method. Moreover, *constr* will use gradients if supplied, otherwise will estimate gradients by finite differences. The major part of the calculation is to solve differential equations, in order to obtain function values, and gradients if used.

Some results from SCOM for the Kendrick-Taylor model (see section 1.4, and section 3.6 for computational details) are presented as follows.

The graphs show:

- The state (Fig. 1) and the control (Fig. 2) for $N = 10$ and 20; for the state, the points x are from $N = 10$.
- In Fig. 3, the effect of increasing the number of subdivisions from $N = 10$ to $N = 20$; for comparison, the points x were obtained using RIOTS_95.

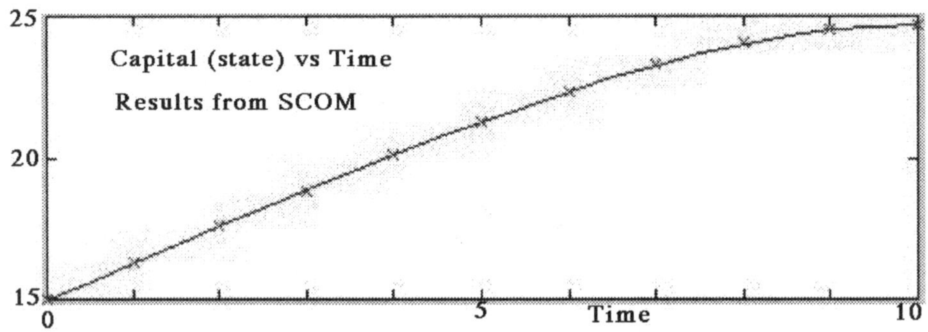

Fig. 1

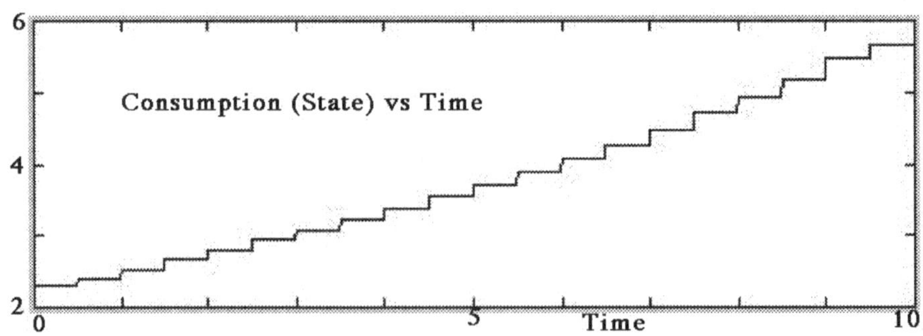

Fig. 2

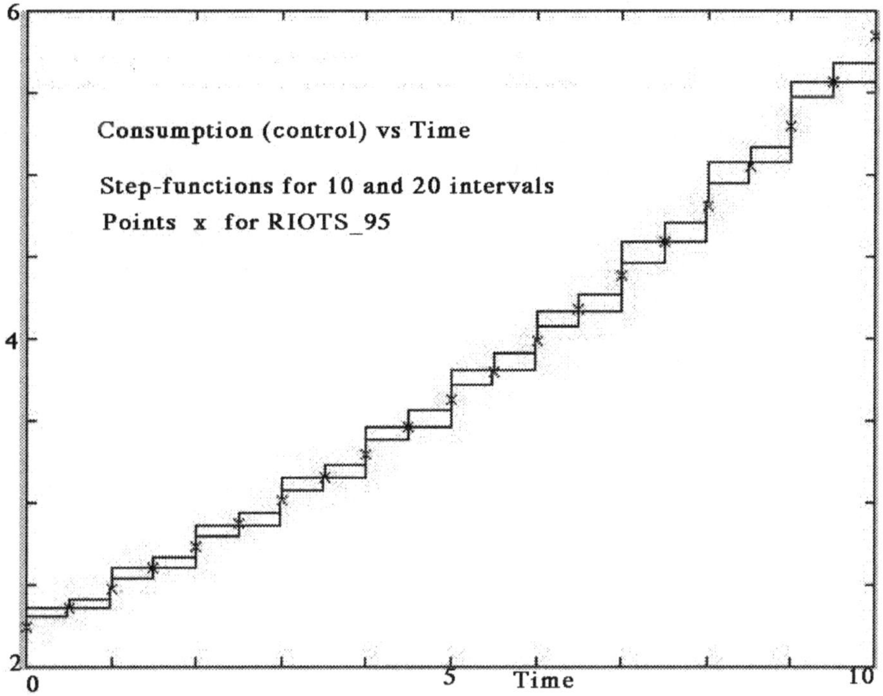

Fig. 3

3.4. Detailed account of the SCOM package

3.4.1. Preamble

SCOM is a computer package, using the *constr* program in MATLAB's Optimization Toolbox, for solving optimal control problems in continuous time. The user must supply a *calling program*, and *user subroutines* for the functions in the problem. There is a choice whether or not gradient subroutines are supplied. To supply them requires more mathematics and coding, gets much faster computation, but does not work with some control problems, if the functions are only defined in a restricted domain. If gradients are not supplied, then *constr* can estimate them, but with less precision.

The MATLAB facilities make experimentation easy, since MATLAB does much of the "housekeeping" (looping, matrix calculations, passing variables) that must be coded explicitly in some other languages.

3.4.2. Format of problem

Optimization is with respect to *state* $x(t)$ and *control* $u(t)$.

$$\mathrm{MIN}_{x(.),u(.)} \int_0^1 f(x(t), u(t), t)dt + \Phi(x(1)) \quad \text{subject to}$$

$$x(0) = x_0,\ (d/dt)x(t) = m(x(t), u(t), t)(0 \leq t \leq 1) \qquad \textit{Dynamic equation}$$

$$a \leq u(t) \leq b, g(u(t)) \leq 0 \quad (0 \leq t \leq 1) \qquad \textit{Control constraints}$$

The time interval is scaled to [0,1]. If there is a constraint $x(1) \leq k$, this is handled by adding a penalty term $\frac{1}{2}\mu[x(t) - k - \mu^{-1}\rho]_{+^2}$ to the objective function; usually ρ is determined by $x(1) = k$.

The control function approximates the control function $u(.)$ by a step-function, dividing [0,1] into *nn* equal subintervals. (Because the dynamic equation has a smoothing effect, set-function controls are usually a sufficient approximation.) Function values (and often also gradients with respect to control) for the objective function are obtained by solving differential equations. They are then supplied to the optimization program *constr* in MATLAB's *Optimization Toolbox.* The computation is considerably faster if gradients are supplied, but this is not suitable for some problems, especially if the functions are not well behaved outside the feasible region of the problem. If gradients are used, then the *adjoint differential equation* is required:

$$(d/dt)\lambda(t) = -(\partial/\partial x)[f(x(t), u(t), t) + \lambda(t)m(x(t), u(t), t)], \lambda(1) = \Phi'(x(1)).$$

If a constraint $x(1) \geq k$ is required then

$$\mu[x(1) - k - \mu^{-1}\rho]_+ = \Phi'(x(1)) + \rho \text{ if } x(1) = k$$

is added to $\lambda(1); \mu$ is a positive penalty parameter.

The user must supply a *calling program* (defining all parameters), and *user subroutines* for the functions of the given control problem.

This use of MATLAB minimizes the amount of programming required for a given control problem, since MATLAB handles matrix calculations and passing of parameters very effectively. (But another programming language may be preferred for a large control problem, if faster computation is needed.)

The acronym *SCOM* stands for *step-function control optimization on Macintosh,* since the intention was to compute optimal control on a desktop computer, rather than on a mainframe or workstation. Note that MATLAB on a Windows computer does the same job.

3.4.3. The SCOM codes - the user does not alter them

The *constr* package may be used in two modes, either with codes supplied for gradients, or without code for gradients, so then *constr* estimates gradients by finite differences. For some problems, the latter is more robust. There are two corresponding versions of *SCOM* for the two cases. For the first case, subroutines *in3f* and *in3g* supply the input function values and gradients to *constr*; test problem 1, in section 3.5, describes this situation. For the second case, the subroutine *intf* supplies function values to *constr*; test problem 2 (an economics model)— describes this case.

```
% Get function values for constr - no gradient
[f,g]=intf(uu,ps,q,xinit)
[a,b,c]=ocf(uu,ps,q,xinit);
f=a; g=b;

% Solve differential equations - no gradients
function [f,g,xm,f0]=ocf(um,ps,q,xinit)
nx=ps(1);nu=ps(2);nn=ps(3);
x0=xinit;
xm=zeros(nx,nn+1); xm(:,1)=x0;
ma=nx;t=0;it=1;hs=1/nn;
px=char(q(1));
xm=fnc(px,nx,nu,nn,xm,ma,t,it,hs,um,xm); %compute state
ma=1;t=0;it=1;hs=1/nn;
zz=zeros(1,nn+1); zz(1)=0; pj=char(q(2));
jm=fnc(pj,nx,nu,nn,zz,ma,t,it,hs,um,xm);%compute integral
xf=xm(:,nn+1) ; pf=char(q(3));
f=jm(nn+1) + feval(pf,xf); %objective
pc=char(q(4));
for ii=1:nn
g(ii)=feval(pc,ii,hs,um,xm); % Control constraint
end
f0=jm(nn+1);f0;

% Organize steps for Runge-Kutta integration - no gradients case
function xk=fnc(pd,nx,nu,nn,fil,ma,t,it,hs,um,xm)
yin=fil(:,1);
xk(:,1)=yin;
while it ¡ nn+1
y=rk4(pd,ma,t,it,hs,yin,nx,nu,nn,um,xm);
xk(:,it+1)=y(1)';
it=y(2);t=y(3); yin=y(1);
end

% Runge-Kutta increments
function y=rk4(pd,ma,t,it,hs,yin,nx,nu,nn,um,xm)
fp=zeros(1,ma);
tt=zeros(1,ma); p=0; q=1;
tz=ic2(pd,ma,t,it,hs,p,q,fp,yin,tt,um,xm);
tt=tz(1); fp=tz(2);
p=0.5;q=2;
t=t+0.5*hs;
tz=ic2(pd,ma,t,it,hs,p,q,fp,yin,tt,um,xm);
tt=tz(1);
fp=tz(2);
tz=ic2(pd,ma,t,it,hs,p,q,fp,yin,tt,um,xm);
```

```
tt=tz(1);
fp=tz(2);
t=t+0.5*hs;p=1;q=1;
tz=ic2(pd,ma,t,it,hs,p,q,fp,yin,tt,um,xm);
tt=tz(1);
=tz(2);
it=it+sign(hs);
y=[yin+tt/6;it;t];

% Runge-Kutta - get function values
function tz=ic2(pd,ma,t,it,hs,p,q,fp,yin,tt,um,xm)
z=yin+p*hs*fp;
ff=feval(pd,t,it,z,yin,hs,um,xm);
fp=ff;
tz=[tt+q*hs*fp;fp] ;

% Forward interpolation
function ix=li3(vv,hh,xi)
nn=length(vv)-1; nz=1/hh;
fr=nz*mod(xi,hh); bb=floor(xi*nz)+1;
if bb < nn, vu=vv(bb+1);
else vu=vv(nn);end
ix=(1-fr)*vv(bb) + fr*vu;

% Backward interpolation (for adjoint DE)
function ix=lil3(vv,hh,xi)
global um xm jm
nn=length(vv)-1; nz=-1/hh;
if xi < 0, xi = 0; end
fr=-nz*mod(xi,hh); bb=ceil(xi*nz)+1;
if bb > 1, vw=vv(bb-1);
else vw=vv(1);
end
ix=(1-fr)*vv(bb) + fr*vw;


% Get function values for constr - gradients supplied
function [f,g]=in3f(uu,par,subs,xinit)
[a1,a2,a3,a4,a5,a6]=ocf4(uu,par,subs,xinit); f=a1; g=a2;

% Get gradients for constr
function [df,dg]=in3g(uu,par,subs,xinit)
[a1,a2,a3,a4,a5,a6]=ocf4(uu,par,subs,xinit);
df=a6; nn=par(3);
pk=char(subs(5));
dg=feval(pk,uu,nn);

% Solve differential equations - gradients supplied
```

```
function [f,g,xm,f0,lm,gr]=ocf4(um,ps,q,xinit)
nx=ps(1);nu=ps(2);nn=ps(3);x0=xinit;
xm=zeros(nx,nn+1); xm(:,1)=x0;
ma=nx;t=0;it=1;hs=1/nn;
px=char(q(1));
xm=fnc(px,nx,nu,nn,xm,ma,t,it,hs,um,xm);
ma=1;t=0;it=1;hs=1/nn;
zz=zeros(1,nn+1); zz(1)=0;
pj=char(q(2));
jm=fnc(pj,nx,nu,nn,zz,ma,t,it,hs,um,xm);
xf=xm(:,nn+1) ;
pf=char(q(3));
f=jm(nn+1) + feval(pf,xf);
pc=char(q(4));
for ii=1:nn
g(ii)=feval(pc,ii,hs,um,xm);
end
f0=jm(nn+1);
ma=nx;t=1;it=nn;hs=-1/nn;pa=char(q(8));pq=char(q(6));
lm(nn+1,:)=feval(pa,nn,xf,um,xm);
lm=fnl3(pq,nx,nu,nn,lm,ma,t,it,hs,um,xm);
hs=1/nn;
pg=char(q(7));
for ii=1:nn
gr(ii,:)=hs*feval(pg,ii,hs,um,xm,lm,nn);
end
```

```
% Organize steps for RK integration - for adjoint DE
function xk=fnl3(pd,nx,nu,nn,fil,ma,t,it,hs,um,xm)
yin=fil(nn+1,:);
xk(nn+1,:)=yin;
while it ¿ 0
y=rk4(pd,ma,t,it,hs,yin,nx,nu,nn,um,xm);
xk(it,:)=y(1);
it=y(2);t=y(3);
yin=y(1);
end
```

Note that linear interpolation of the state is required in integrating the objective function, and in solving the adjoint differential equation. The latter is solved backwards in tine t, so needs backwards interpolation.

3.5. The first test problem

The following example has two controls and one state, and gradients are calculated.

$$\text{MIN} \int_0^1 [(u_1(t) - 1)x(t) + 0.25u_2(t)]dt \text{ subject to}$$

$$x(0) = 0.5, (d/dt)x(t) = x(t)u_1(t) + u_2(t),$$

$$0 \leq u_1 \text{ (t)}, 0 \leq u_2(t), \ u_1(t) + u_2(t) - 1 \leq 0 \ \ (0 \leq t \leq 1).$$

The adjoint differential equation is:

$$(d/dt)\lambda(t) = -(1 + \lambda(t))u_1(t) + 1, \lambda(1) = 0$$

The gradient of the objective with respect to the controls is obtained by integrating $[\lambda(t)x(t) + x(t), \lambda(t) + 0.25]$; this is the gradient of the Hamiltonian:

$$(u_1(t) - 1)x(t) + 0.25u_2(t) + \lambda(t)[x(t)u_1(t) + u_2(t)]$$

with respect to the controls.

Note that the constraints on the control are handled separately, since the optimizer *constr* handles such constraints. This example has a switching time of exactly $t = 0.75$. This is only approximated with $N = 10$ subdivisions; some other initial control u gave a positive $u(t)$ in subinterval 8. While $N = 20$ subintervals happens to give an exact solution, in general the switching time would have to be made a parameter, to be optimized over.

Calling program for first test problem

```
%run3.m
format compact
subs=cell(1,9);
subs='t3x','t3j','t3f','t3c','t3k','t3l','t3g','t3a';
par = [1 2 10];
xinit=[0.5];
u0 = zeros(10,2); ul = zeros(10,2); uu = ones(10,2);
Control=constr('in3f',u0,[],ul,uu,'in3g',par,subs, xinit)
[Obj, Con,State,Int,Costate,ObjGrad]=ocf4(Control,par,subs,xinit);
State,Obj,Int,Costate,ObjGrad
```

For this problem, the *subs* line lists also the subroutines (here *t3k.m, t3l.m, 63g.m, t3a.m* describing gradient of constraint, right hand side of adjoint differential equation, gradient of objective (with respect to control), and costate boundary condition (at time $t = 1$.)

% The following lines plot one state component against another, and plot two control components, which are step-functions, against time

```
plot(State(:,1), State(:,2),'x-')
xlabel('State(1')
ylabel('State(2')
figure
t2=[0:0.001:0.999];
plot(t2, Control(floor(nn*t2)+1,1),'r')
hold on
plot(t2, Control(floor(nn*t2)+1,2),'g')
```

The function subroutines are as follows:

```
% Right hand side of differential equation
function yy=t3x(t,it,z,yin,hs,um,xm)
yy(1)=z(1)*um(floor(it),1)+um(floor(it),2);

% Integrand of objective function
function ff=t3j(t,it,z,yin,hs,um,xm)
ff(1)=(um(floor(it),1)-1)*li3(xm,hs,t);
ff(1)=ff(1)+0.25*um(floor(it),2);

% Objective endpoint term
function ff=t3f(xf,um,xm)
ff(1)=0;

% Control constraint
function gg=t3c(ii,hs,um,xm)
gg=um(ii,1) + um(ii,1) - 1;

% Gradient of constraint
function dg=t3k(um,nn)
dg=[eye(nn);eye(nn)];

% RHS of adjoint equation
function yy=t3l(t,it,z,yin,hs,um,xm)
yy=-(1+z(1))*um(floor(it),1) +1;

% Gradient of objective
function yy=t3g(t,hs,um,xm,lm,nn)
temp= 0.5*(lm(t,1)+lm(t+1,1)); t2=t/nn;
yy=[(1+temp)*li3(xm,hs,t2), 0.25+temp];

% Costate boundary condition
function yy=t3a(nn,xf,um,xm)
yy=0;
```

3.6. The second test problem

This version of the Kendrick-Taylor model for economic growth (see Kendrick and Taylor, 1971) has one state and one control. This problem has implicit constraints, and a computation with gradient formulas supplied failed. (The dynamic equation ran $x(t)$ into negative values, yet fractional powers of x(t) were required.) An optimum was computed using *constr* to estimate gradients by finite differences.

With capital stock at time t as state function, and rate of consumption at time t as the control function, this model has the form:

$$MAX \int_0^T e^{-\rho t} u(t)^{\tau} dt \text{ subject to}$$

$$x(0) = k_0, \ \dot{x}(t) = \zeta e^{\delta t} x(t)^{\beta} - \sigma x(t) - u(t), \ \ x(T) = k_T.$$

To allow comparison with numerical results in Kendrick and Taylor (1971), the following numerical values were used:

$$T = 10, \tau = 0.1, k_0 = 15.0, \zeta = 0.842, \beta = 0.6, \sigma = 0.05, k_T = 24.7.$$

Although this model states no explicit bounds for $x(t)$ and $u(t)$, both the formulas and their interpretation requires that both $x(t)$ and $u(t)$ remain positive. But with some values of $u(.)$, the differential equation for $\dot{x}(t)$ can bring $x(t)$ down to zero; so there are implicit constraints.

This example was computed as a minimization problem, with $N = 10$ and $N = 20$ subdivisions (the latter required reduced termination tolerance and more iterations). The different computations differed very little in the optimum objective value. Evidently, the optimum objective here is fairly insensitive to small changes in the control. The convergence to an optimum objective value is fast, but the convergence to an optimal control is slow. There is some scope here for adjusting, within *constr,* the termination tolerances for gradient and for satisfying constraints. Because of a difficulty with the fractional powers, this problem was computed with gradients obtained by *constr* from finite differences, instead of using gradients from the costate equation.

Calling program for second test problem

```
%kt2run.m
format compact
subs=cell(1,4); subs='kx','kj','kf','kc';
par=[1 1 10 15]; % 1 state, 1 control, 10 subintervale,
ul=ones(10,1); % control lower bound is 1.0
uh=10*ones(10,1); % control upper bound is 10.0
xinit=[15]; % Initial state
u0=ones(1,10);
Control=constr('intf',u0,[],ul,uh,[],par,subs,xinit)
[Objective, Constraint, State, Integral]=ocf(Control,par,subs,xinit);
Objective, State, Integral
```

Note that the subroutines (kx.m, kj,m, kf.m, kc.m) for the functions specifying the problem, namely the right hand side of the differential equation, the integrand and the endpoint term of the (minimizing) objective function, and the control constraint (absent here) are listed in the *subs* line. These subroutines are specific to tbe particular problem being solved. Upper and lower bounds for the control are specified by the vectors *uu* and *ul*; they may depend on time. The initial condition for the state is given as *xinit*. The vector *par* may have additional parameters added, if required, to be passed to the function subroutines.

```
% Right hand side of differential equation
function yy=kx(t,it,z,yin,hs,um,xm)
yy=0.842*exp(0.2*t)*z(1)∧0.6 - um(floor(it)) - 0.05*z(1); yy=10*yy;

% Integrand of objective function
function ff=kj(t,it,z,yin,hs,um,xm)
ff=-10*exp(-0.3*t)*um(floor(it)∧0.1;

% Objective endpoint term
function ff=kf(xf,um,xm)
ff=10*(xf-24.7)∧2;

% Control constraint
function gg=kc(ii,hs,um,xm)
gg=0;
```

3.7. The third test problem

The Davis-Elzinga Investment Model (Davis and Elzinga, 1972) considers the optimal policy for dividends and stock issues for a utility company.

The model:

At time t, $P(t)$ is the market price of a share of stock, and $E(t)$ is the equity per share of outstanding common stock. This leads to an optimal control model:

$$\text{MAX} \int_0^1 e^{-\rho t}(1 - u_r\ )\text{rETdt} \ - e^{-\rho T}P(T)$$

subject to dynamic equations:

$$\dot{P} = cT[)1 - u_r)rE - \rho P]\ ,\ \dot{E} = rTE[u_r + u_s(1 - E/((10\delta)P)]\ ,$$

and initial conditions $P(0)$ and $E(0)$, and constraints $u_r \geq 0$, $u_s \geq 0$,

$$u_r + u_s \leq \text{ b.}$$

The multiplier T is required to convert the time scale from [0, T] to [0,1] for computation. The computation calls MATLAB's *constr* package for constrained minimization.

Calling program

```
% dn2_run.m The calling program, setting the parameters
subs=cell(1,9);
subs={'dnx','dnj','dnf','dnc'};
par=[2, 2, 20, 0, 0, 1, 6, 0.2, 0.1, 0.1, 0.75]; % Parameters
% nx, nu, nn, npa, grad, c, T, r, p, d, b=k/r
xinit=[2 1]; nn=par(3);
u0=zeros(nn,2);
ul=zeros(nn,2);
uu=ones(nn,2); figure
Control=constr('fqq',u0,[],ul,uu,[],par,subs,xinit)
[Objective,Constraint,State,Integral]= cqq(Control,par,subs,xinit)
plot(State(:,1),State(:,2),'x-') % Graph outputs
xlabel('State 1')
ylabel('State 2')
figure
t2=[0:.001:0.999];
plot(t2,Control(floor(nn*t2)+1,1),'r')
hold on
plot(t2,Control(floor(nn*t2)+1,2),'g')
```

User subroutines

```
function ff= dnx(t,it,z,yin,hs,um,xm,ps)
                              % The right hand sides of the DEs
c=ps(6);T=ps(7);r=ps(8);p=ps(9);d=ps(10);b=ps(11);
ff(1)=c*T*((1-um(floor(it),1))*r*z(2)-p*z(1));
temp=um(floor(it),2)*(1-z(2)/((1-d)*z(1)));
ff(2)=r*T*z(2)*(um(floor(it),1)+temp);

function ff= dnj(t,it,z,yin,hs,um,xm,ps) % The integrand
c=ps(6);T=ps(7);r=ps(8);p=ps(9);d=ps(10);b=ps(11);
temp=iqq(xm,hs,t);ww=temp(2);
ff(1)=-(1-um(floor(it),1))*r*T*ww*exp(-p*T*t);

function gg= dnc(ii,hs,um,xm,ps) % The constraints on the control
c=ps(6);T=ps(7);r=ps(8);p=ps(9);d=ps(10);b=ps(11);
gg(1) = um(floor(ii),1)+um(floor(ii),2)-b;

function ff= dnf(xf,um,xm,ps) % Endpoint term of objective
c=ps(6);T=ps(7);r=ps(8);p=ps(9);d=ps(10);b=ps(11);
ff(1)=-xf(1)*exp(-p*T);
```

Computed output, showing the effect of initial conditions on P and E

Note that the theoretical solution is *bang-bang control* , thus the control vector (u_r, u_s) switches between three régimes $A = (0,0), B = (b,0), C = (0,b)$.), which are the vertices of the triangular region to which (u_r, u_s) is restricted. From the theory, the optimum switches from B to A if $P(T)/E(T) <$

1.467, or from B to C if $P(T)/E(T) > 1.467$. The switching may not be observed unless the planning horizon T is large enough.

Three computed results are presented, with different initial conditions for the states P and E.

Initial state is [0.2 1]

Control =
0.7500 0
0.7500 0
0 0
0 0
0 0
0 0
0 0
0 0
0 0
0 0
0 0
0 0
0 0
0 0
0 0
0 0
0 0
0 0
0 0
0 0

Thus régime B, then A

Objective = 1.4615

State =
0.2000 1.0000
0.2092 1.0460
0.2188 1.0942
0.2770 1.0942
0.3335 1.0942
0.3883 1.0942
0.4415 1.0942
0.4932 1.0942
0.5433 1.0942
0.5919 1.0942
0.6391 1.0942
0.6849 1.0942
0.7293 1.0942
0.7724 1.0942
0.8143 1.0942

0.8549 .0942
0.8943 .0942
0.9325 1.0942
0.9696 1.0942
1.0057 1.0942
1.0406 1.0942
Integral = 0.8904; P(T)/E(T)=0.952

Initial state is [0.5 1]
Control =
0.7500 0
0.7500 0
0 0
0 0
0 0
0 0
0 0
0 0
0 0
0 0
0 0
0 0
0 0
0 0
0 0
0 0
0 0
0 0
0 0
0 0.0294
Thus régime B , then A
Objective = 1.5518
State =
0.5000 1.0000
0.5003 1.0460
0.5014 1.0942
0.5512 1.0942
0.5996 1.0942
0.6466 1.0942
0.6921 1.0942
0.7363 1.0942
0.7793 1.0942
0.8209 1.0942
0.8613 1.0942
0.9005 1.0942

0.9386 1.0942
0.9755 1.0942
1.0114 1.0942
1.0462 1.0942
1.0799 1.0942
1.1127 1.0942
1.1445 1.0942
1.1753 1.0942
1.2053 1.0941
Integral = 0.8904; P(T)/E(T)=1.112

Initial state is [2 1]
Control =
-0.0000 0.7500
0.0000 0.7500
0 0.7500
0 0.7500
0 0.7500
0 0.7500
-0.0000 0.7500
0 0.7500
-0.0000 0.7500
0 0.7500
0 0.7500
0 0.7500
0 0.7500
-0.0000 0.7500
0 0.7500
-0.0000 0.7500
-0.0000 0.7500
-0.0000 0.7500
-0.0000 0.7500
-0.0000 0.7500
Thus régime C always
Objective = 2.2621
State =
2.0000 1.0000
2.0006 1.0199
2.0023 1.0398
2.0052 1.0595
2.0092 1.0792
2.0142 1.0987
2.0201 1.1181
2.0271 1.1374
2.0350 1.1567

2.0438	1.1758
2.0535	1.1948
2.0640	1.2138
2.0753	1.2327
2.0874	1.2515
2.1002	1.2703
2.1138	1.2890
2.1281	1.3077
2.1430	1.3264
2.1586	1.3450
2.1749	1.3636
2.1918	1.3822

Integral =1.0592; P(T)/E(T)=1.586

Régime B, then A   Régime B then A   Régime C always

(From theory, there are two régimes, depending on whether $P(T)/E(T) <$ or > 1.467 ; so the prediction is:

B then A,   B then A   B then C (if T is large enough)

Chapter 4
Computing Optimal Growth and Development Models

4.1. Introduction

Methods and models for optimal growth and development are often formulated as optimal control problems. They can represent a classical example of a social choice problem in an economy. Such models may describe intertemporal allocation of resources, so as to maximize social welfare, determined on the basis of the underlying social preferences and value judgements. They may describe a national economy, with state functions being vector valued, since various commodities are involved. The consumption may be varied, within some limits, to optimise some utility functional, which may involve both state and control functions over a planning period. In these models, there are in general explicit bounds (such as floor to consumption, or a ceiling to capital or pollution), as well as discontinuous control functions. These social choice models for optimal growth and development can be specified, and mumerically implemented, following the assumptions of new^3 welfare economics, discussed in chapter 1.

Analytic solutions are available for some of the simpler models. Often some sort of steady state has been studied, but not the rate of approach to the steady state from given initial conditions, or the transition to give terminal conditions (such as, for example, a stated minimum for terminal capital, so that a process may continue). It is noted that the available data are necessarily in discrete time (i.e. annual or quarterly data). While discrete-time optimal control models can of course be computed, a continuous-time model may give a more intelligible picture of what is happening. Also, the Pontryagin theory for optimal control applies to optimal control in continuous time, but only under serious restrictions to optimal control in discrete time. This chapter will hence consider continuous-time models.

A computer package allows exploration of sensitivity to parameter changes, and of the domains of the various parameters for which optimal solutions exist. Some recently developed computer packages, coding different algorithms for continuous optimal control, such as *RIOTS, MISER* and *OCIM,* (which use MATLAB), have current applications to other areas of control, but have not yet been used for computing optimal growth models. The objective of this chapter is to demonstrate the applications of these algorithms and computer programs to the computation of optimal growth models. The MATLAB package *SCOM*, described in chapter 3, is used to study the Kendrick and Taylor model (Kendrick & Taylor, 1971). The computational results are consistent with the economic conclusions obtained by these authors, and by Chakravarty

(1969), and also with results computed with another package. With this computational simulation, it is easy to study the sensitivity of the results to changes in model parameters.

In addition, the optimum obtained has been proved to be the global optimum, using invexity properties (see section 2.4). The global optimality of the model results is also proved in this chapter.

The chapter is organised as follows. Section 2 presents a Ramsey type (Ramsey, 1928), optimal growth model developed by Kendrick and Taylor (1971) and its optimal control specification. Section 3 discusses different algorithms and computer programs for solving optimal control models, while Section 4 discusses the Kendrick-Taylor model and numerical implementation by some computer programs. Mathematical properties of the model results and the sensitivity studies are reported in Section 5 respectively. Section 6 refers to the experiences for computing optimal growth models by computer programs, such as *OCIM* and *MISER*. Section 7 contains the conclusion.

4.2. The Kendrick-Taylor Growth Model[7]

The well known Kendrick-Taylor model for economic growth and development (Kendrick and Taylor, 1971) has been used as a test problem for computational approaches. With capital stock k(t) at time t as the state function, and rate of consumption c(t) at time t as the control function, this model has the form:

$$\text{MAX} \int_0^T e^{-\rho t} c(t)^{\tau} dt \quad \text{subject to:}$$

$$k(0) = k_0, \quad \dot{k}(t) = \zeta e^{qt} k(t)^{\beta} - \sigma k(t) - c(t), \quad k(T) = k_T.$$

This model does not state any explicit bounds for $k(t)$ and $c(t)$. However, both the formulas and their interpretation requires that both $k(t)$ and $c(t)$ remain positive. However, with some values of $u(t)$, the differential equation for $k(t)$ can bring $k(t)$ down to zero.

Consistent with the mainstream practice in modelling optimal growth, development and welfare economics, the numerical optimal growth programs in this chapter are specified within the framework of the elements of an optimal growth program of the following form: an optimality criterion contained in an objective function which consists of the discounted sums of the utilities provided by consumption at every period; the finite planning horizon; a positive discount rate; and the boundary conditions given by the initial values of the variables and parameters and by the terminal conditions.

This model can be expressed by a standard kind of optimal control model, which may be written as follows:

$$\text{MIN}_{k(.),c(.)} \int_0^T f(k(t), c(t), t) dt + \Phi(k(T)) \quad \text{subject to}$$

[7] See also section 2.1.

$$k(0) = k_0, \dot{k}(t) = m(k(t), c(t), t), k_L(t) \leq k(t) \leq k_U(t) \ (0 \leq t \leq T).$$

Here:

$$f(k(t), c(t), t) = -e^{-\rho t}c(t);\ \Phi(k(T)) = \mu(k(T) - k^*)^2,$$

(with a minus sign on f to convert to a minimization problem), and:

$$m(k(t), \varsigma(t), t) = \zeta e^{\delta t} k(t)^\beta - \sigma k(t) - c(t).$$

The terminal constraint for $k(T)$ has been replaced by a penalty cost term $\Phi(k(T))$, with a parameter k^* (approximately k_T, but may need adjustment), and a sufficiently large coefficient μ. In general (though not for the present instance), $k(t)$ and $c(t)$ are vector valued.

The relevant mathematical issues that are investigated in the context of an optimal growth model are the following: existence, uniqueness and globality of the optimal policy solution, and stability of the dynamic system in the equilibrium or steady state position (Intriligator 1971).

The control issues of analytical and policy importance in control models are stated by Sengupta and Fanchon (1997) as: estimability, controllability, reachability, and observability. As these characteristics of control models relate to models with linear dynamics, but not to nonlinear systems, they are not investigated here. However, some numerical data relating to controllability are given in Section 4.4.

The Kendrick-Taylor model was numerically implemented by Kendrick and Taylor (1971) by several algorithms such as search and quasi- linearisation methods based on the discrete Pontryagin maximum principle. In another experiment, Keller and Sengupta (1974) solved the model by conjugate search and Davidson algorithm based on the continuous Pontryagin maximum principle. While these algorithms were implemented by some special purpose computer programs, a GAMS version (a general purpose commercially available program) of the Kendrick-Taylor model is also available.

4.3. The Kendrick-Taylor Model Implementation

To allow comparison with numerical results in Kendrick and Taylor (1971), the following numerical values were used:

$$T = 10,\ \tau = 0.1,\ \ \zeta = 0.842,\ \ \beta = 0.6,\ \ \sigma = 0.05,\ \ k_T = b = 24.7.$$

The Kendrick-Taylor model, with the parameter values listed above, has been computed using the RIOTS_95 package on a Pentium, and also with some of the parameters varied, as listed below. For comparison, the model has also been computed using the SCOM package on an iMac computer. The latter package also uses MATLAB (version 5.2) and the constrained optimisation solver *constr* from MATLAB's Optimisation Toolbox, but none of the differential equation software used by RIOTS_95. Thus both computations share

MATLAB's basic arithmetic and display software, but the implementations are otherwise independent.

The result of the computed optimum growth paths of aggregate variables and ratios (the great ratios of economics (see Fox et al., 1973)) that show the empirical process of growth and trends of the economy are reported in Figure 1. The Kendrick-Taylor model was solved for 10 periods, for a total time of 10 years, using RIOTS_95.

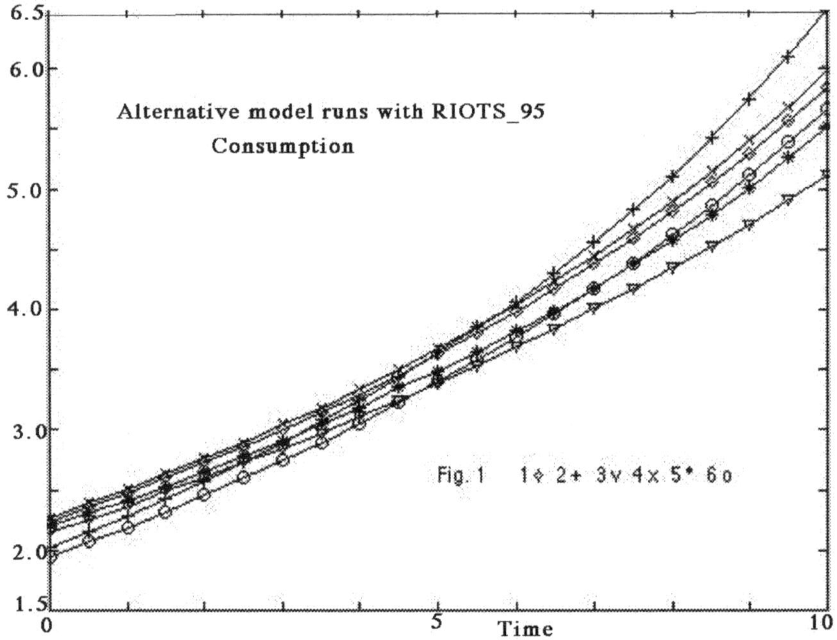

Figure 1a. Alternative Model Runs with RIOTS : Consumption

For sensitivity analysis, the following set of alternative values of the parameters of the model was adopted in six model runs, shown in the following Table.

Run	ρ	δ	k_T	ζ
1	0.03	0.02	24.7	0.842
2	0.01	0.02	24.7	0.842
3	0.03	0.01	24.7	0.842
4	0.03	0.02	23.7	0.842
5	0.03	0.02	24.7	0.822
6	0.01	0.01	24.7	0.842

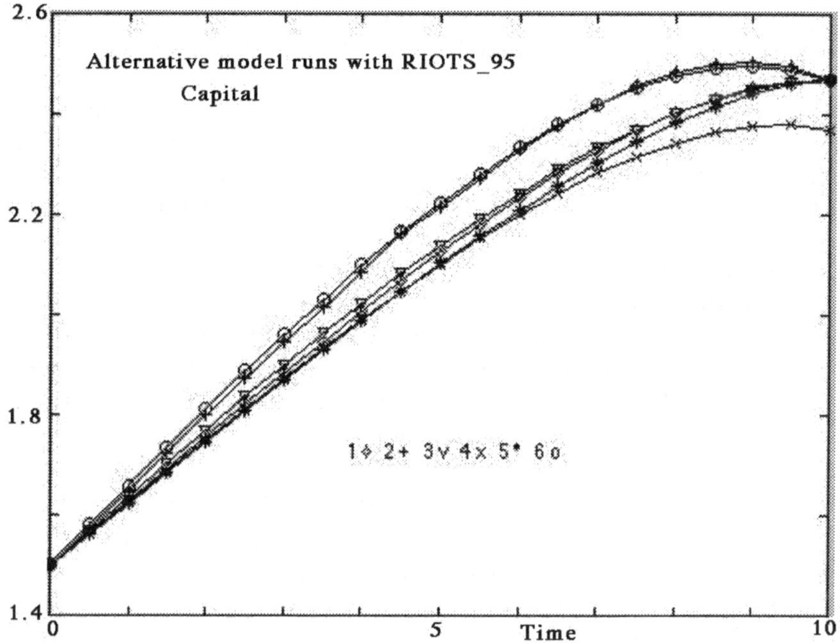

Figure 1b. Alternative Model Runs with RIOTS : Capital

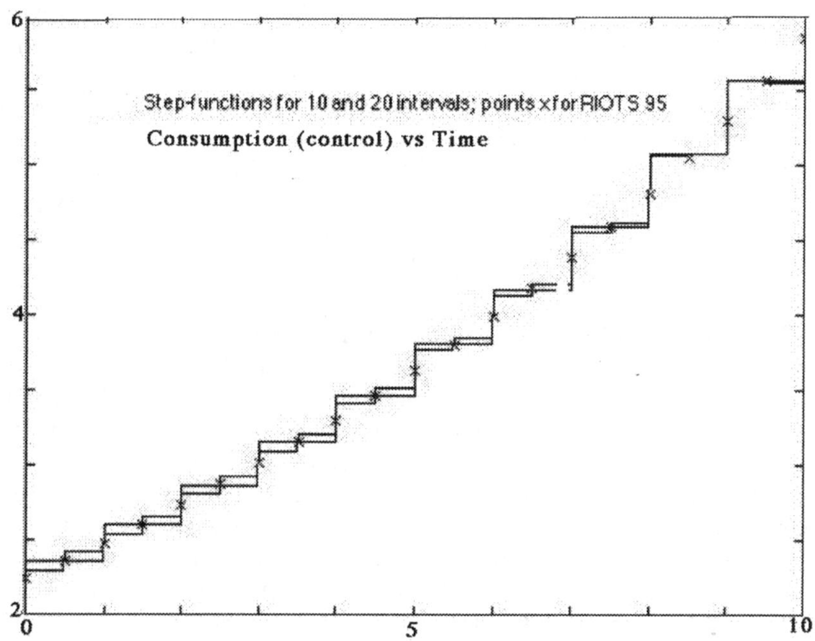

Figure 2. Effect of Number of Subdivisions

These results of the Kendrick-Taylor model show the following numerical characteristics of optimum growth of the economy: (1) the growth paths of the major factors that determine growth rate and process of the economy; (2) an optimum composition of GDP consisting of consumption, saving and investment; (3) the evolution of dynamic rental price path that satisfies the given historical conditions and ensures the attainment of terminal capital stock; (4) a growth rate of the economy that provides a net rate of return of the economy equal to the social time preference rate; and (5) the growth path of the economy that generate highest level of social welfare for the economy, obtained with the three computations.

Figure 2 shows the control (consumption) for N = 10, compared with N = 20, as computed by SCOM. For comparison, the control points shown x were obtained with RIOTS_95. The state is the same, to graphical accuracy, in the three computations.

4.4. Mathematical and Economic Properties of the Results

The results of these models describe the optimal social choice regarding intertemporal allocation of resources, for an economy satisfying the underlying social preferences and value judgements embedded in the model specification and implementation.

The significant properties of the optimally growing economy characterised by the trajectories of the Kendrick-Taylor model computed here numerically are that the economy grows along a unique equilibrium path, and along this path the society's saving (which is equal to investment) and consumption are optimal and the work force is employed at the full employment level. The values of the costate variable in the Kendrick-Taylor model at different time periods are the shadow prices along the dynamic equilibrium path of the economy and provide the dynamic valuation or pricing system of the economy.

Mathematical properties such as existence, uniqueness, stability and other properties of a steady state solution of the Kendrick-Taylor model are important in deriving and understanding the above stated economic implications of the growth model.

The first issue is the existence of an optimal growth trajectory of the model. Consider first a discrete-time version of the Kendrick-Taylor model, in which the control u is a n-dimensional vector. Suppose that lower and upper bounds are imposed on $u(.)$. Then the set $U(n)$ of feasible controls is bounded, as well as closed; hence the objective function, being continuous, reaches its extremum at a point of $U(n)$ thus, the optimum is attained. Now computing an optimum for the continuous-time version of the model involves the tacit assumption that its optimum is a limit of the optima for suitable finite-dimensional approximations. Assuming this, the attainment of the continuous-time optimum would follow as a limiting case. However, the attainment and uniqueness of the optimum can be deduced in another way. The Pontryagin theory gives first-order necessary conditions for an optimum, and these conditions have a solution. If the model were a convex problem, then the first order necessary conditions would also be sufficient for an optimum, thus proving its existence. But convexity is lacking, in particular because the equality constraint (the differential equation) is not linear. However, it is shown below that there is a transformation of the model into a convex problem; hence the solution is a global optimum.

Is there a transformation of the variables $k(t), c(t)$ to some new variables $x(t), u(t)$, so that the transformed problem is a convex problem? If there is such a transformation, then the original problem has a property called invex (Craven 1995); and it follows then that a local optimum is also a global optimum. The following calculation shows that the Kendrick-Taylor model, with the parameters stated above (from Kendrick and Taylor, 1971), is transformable to a convex problem. This will also hold with various other values of the parameters. For this purpose, the Kendrick-Taylor problem may be considered as a minimisation problem, to minimise the negative of the given objective function.

First define $K(t) := k(t)e^{\sigma t}$. Then:

$$\dot{K}(t) = \zeta e^{rt} K(t)^{\beta} - e^{\sigma t} c(t),$$

where $r = \delta + (1-\beta)\sigma$. This gets rid of the the $\sigma k(t)$ term in the differential equation.

Next, define a new state function $x(t) = K(t)^{\gamma}$, where $\gamma = 1 - \beta$ will be assumed later. Denote also $\theta = (1-\gamma)/\gamma$. Then:

$$\dot{x}(t) = \gamma \zeta e^{rt} x(t)^{(\beta+\gamma-1)/\gamma} - \gamma e^{\sigma t} x(t)^{-\theta} c(t)$$

The further substitution $c(t) = x(t)^{\theta} u(t)$ reduces the differential equation to the linear form :

$$\dot{x}(t) = \gamma \zeta e^{rt} - \gamma e^{\sigma t} u(t)$$

What becomes of the integrand $-e^{-\rho t} c(t)^{\tau}$ in the objective function? It becomes $-e^{-\rho t} x(t)^{\kappa} u(t)^{\tau}$, where $\kappa = \theta\tau$. Note that with the numerical values cited for τ and θ, $\kappa = 0.15$, and $\kappa + \tau < 1$. Since $c(t)$ and $k(t)$ must be positive, the domain of $(x(t), u(t)$ is a subset of $\mathbf{R}^2_+$, depending however on t. The Hessian matrix of $-x^{\kappa} u^{\tau}$ is then:

$$\begin{pmatrix} -\kappa(\kappa-1)x^{\kappa-2}u^{\tau} & -\kappa\tau x^{\kappa-1}u^{\tau-1} \\ -\kappa\tau x^{\kappa-1}u^{\tau-1} & -\tau(\tau-1)x^{\kappa}u^{\tau-2} \end{pmatrix}$$

Since $0 < \kappa < 1$ and $0 < \tau < 1$, the diagonal elements of the Hessian are positive. The determinant is calculated as:

$$x^{2\kappa-2} u^{2\tau-2} \kappa\tau(1-\kappa-\tau)$$

which is positive if $\kappa + \tau < 1$.

Values $\lambda = 0.1$ and $\beta = 0.6$ have been used in some economics references. Invexity, and consequent global optimization, is thus shown for these values, and also for other values near them. Thus, global optimization still holds when the exponents β and q and the decay coefficient σ differ substantially from the values computed here.

The optimum curves do not, however, approach a steady state as time increases, because the model contains the exponential terms $e^{-\rho t}$ and e^{qt}, and the coefficient σ which leads to an exponential term $e^{\sigma t}$. The optimum curves are, however, fairly insensitive to a small change in the endpoint condition $k(t) = k_T$.

For the dynamic systems, stability of the system can be determined by the characteristic roots of the Jacobian matrix formed by linearizing the differential equations of the variables and evaluated at the steady state point.

The results in Figure 1 show the RIOTS results of how changes in ρ, δ, k_T, and ζ affect the optimum growth trajectories. Figure 1 shows that changes in ρ and k_T (terminal capital) have relatively higher impact on the dynamic path of capital accumulation compared to the changes in the rate of technical progress and the elasticity of marginal utility with respect to changes in income. In regard to the value of optimal control (the optimal level of consumption), the impacts are similar. The optimal control variable is relatively more sensitive (see Figure 1a) to changes in the discount rate and the terminal level of capital. These sensitivity experiments suggest that the social time preference and the terminal conditions are significant determinants of the structure of optimal growth paths of an economy.

RIOTS was also used to examine the sensitivity of the Kendrick/Taylor model to some other changes in the data. Figure 3 shows the trajectory of the capital when the initial capital is varied from 15.0 to 14.0 or 16.0, though not changing the terminal constraint on the capital. Figure 4 shows the trajectory of the capital with the initial value fixed at 15.0, but the terminal value varied from 24.7 to 27.0 and 22.4. The time scale here is 0 to 10; the turnpike effect noted by Kendrick and Taylor is only apparent on a longer time scale, say 0 to 50. Figure 5 shows the effect of varying the exponent β in the dynamic equation from its given value of 0.6. The upper curve has $\beta = 0.5$, compared with the middle curve for $\beta = .6$. The lower curve is what RIOTS gives when $\beta = 0.7$ but with the warning that no feasible point was found; a computation with SCOM confirms that the model is not feasible for this case.) This graph is included, to point out that when the constraints cannot be satisfied, a computer output need not be any sort of approximation to a solution. In this instance, an implicit constraint is violated, since both state and control run negative.

RIOTS computed an optimal solution to the Kendrick-Taylor model in 20 iterations. The speed is due in part to the user by RIOTS of C code - although the functions defining the problem were entered as M-files (requiring however some computer jargon for switching). SCOM was slower - fairly rapid convergence to the optimum objective value, but much slower convergence to the optimal control function. Some more development is needed here with terminating tolerances for gradient and constraints.

The computing experience in this chapter is largely similar to the previous computing experiences with the Kendrick-Taylor model (Kendrick and Taylor (1971)).

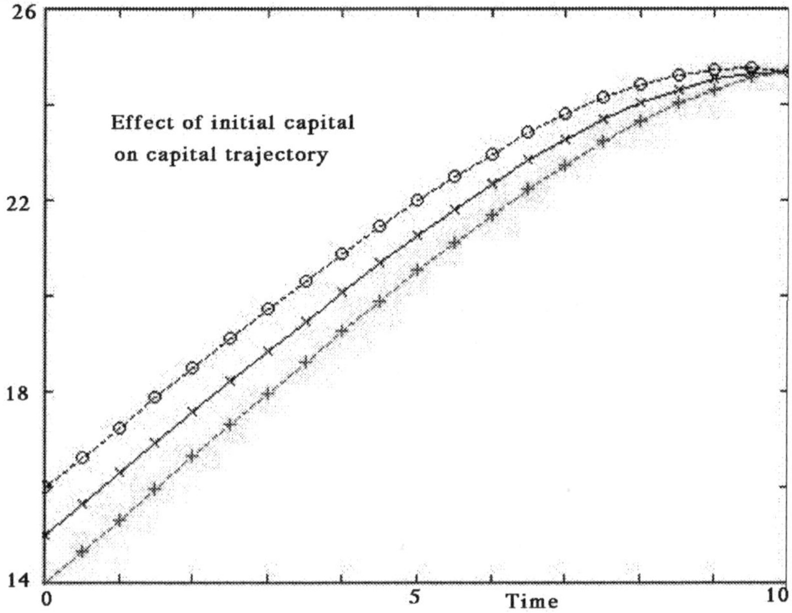

Fig. 3 Effect of initial capital

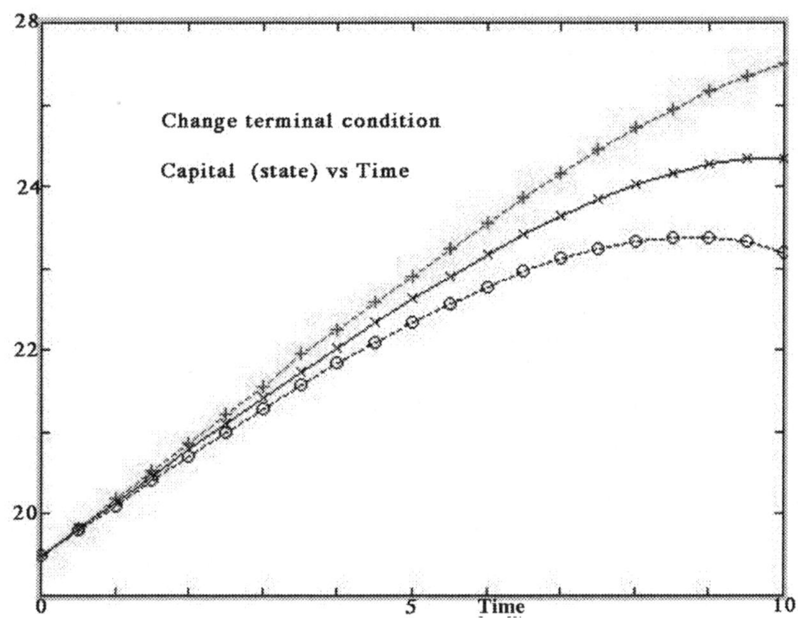

Fig. 4 Effect of terminal condition

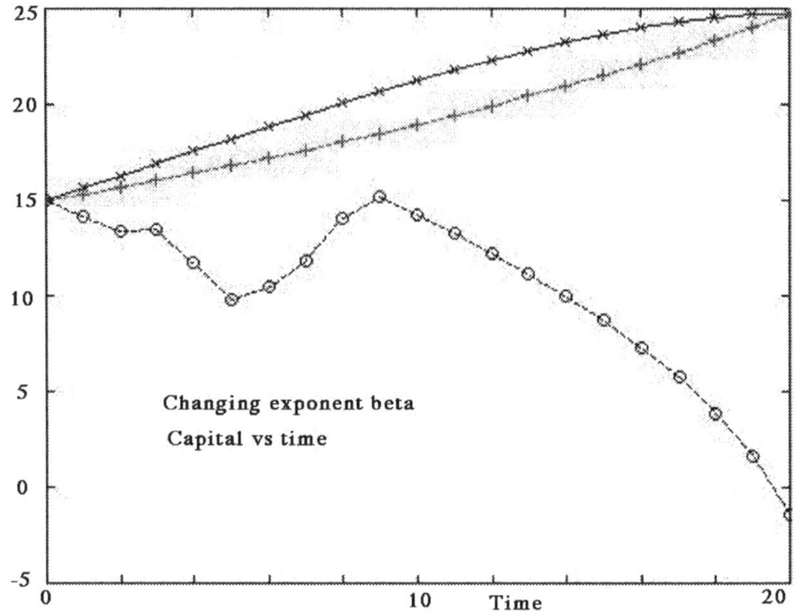

Fig. 5 Effect of changing exponent beta

4.5. Computation by other computer programs

While the Kendrick-Taylor model was not computed with MISER, some related economic models were attempted, without success because sufficient accuracy was not obtained for the gradients of the functions. The experience with SCOM suggests that the fractional-power terms in the model may lead to computational difficulty, since some values for the control function may lead to negative values for the state function, for which the fractional powers are undefined.

4.6. Conclusions

Algorithms and computer packages for solving a class of optimal control problems in continuous time, in the form of optimal growth, development and finance models (see chapters 3, 4 and 5), using the MATLAB system, but in a different way from the RIOTS_95 package which also uses MATLAB and the RIOTS system, have produced plausible economic results. In the SCOM approach as in the MISER and OCIM packages, the control is parameterised as a step-function, and SCOM uses MATLAB's *constr* package for constrained optimization as a subroutine. End-point conditions are simply handled using penalty terms. Much programming is made unnecessary by the matrix features built into MATLAB. Some economic models present computational difficulties because of implicit constraints, and there is some advantage using finite difference approximations for gradients. The RIOTS system has the discretisation approach. It can produce however a set of results close to these produced by the SCOM approach. While several computer packages are available for optimal control problems, they are not always suitable for a wide class of control problems. The MATLAB based RIOTS and SCOM, however, offer good opportunities for computing continuous optimal-growth models. Optimal growth modellers may find these algorithms and computer programs suitable to their work.

Chapter 5

Modelling Financial Investment with Growth

5.1. Introduction

There are many models for financial investment, assuming that a productive economy is available to invest in. There are many economic models, commonly assuming that investment is a given quantity. However, questions arise as to the mutual interaction of these two areas, especially the *positive feedback* which may drive price changes. This chapter seeks to model some aspects of this interaction, without assuming that the financial investors or planners always behave (or should behave) according to optimal economic principles. These models, like the other models in this book, assume new[3] economics, and may be applied to *social choice* for an optimal intertemporal allocation of financial and physical resources in the economy.

Should such a model be in discrete or continuous time? Models in continuous time usually assume a Brownian motion as the basis of the stochastic contribution, carrying with it a questionable assumption of independent contributions from non-overlapping time intervals, however short. A discrete- time model assumes less, and relates better to data which are available at discrete times. Hence a discrete-time model is adopted here, including a stochastic increment at each discrete time. If these increments are not too large, the results can be adequately described by the evolution of the mean (or expectation) and the variance. Thus, more complicated calculations about probability distributions may be avoided.

When an objective function is to be maximized, for such a model, it is expedient to approximate the dynamic difference equations for the mean and variance by differential equations, in order to use available computer packages for optimal control. However, the results still relate to the discrete- time model.

The present approach makes some use of recent mathematics in optimization, which has been applied to some economic models (see e.g. Islam, 2001a; Chakravarty, 1979; Leonard and Long, 1992))

Some further discussion is given of multiple steady states in such dynamic models, and also of sensitivity questions, especially when the planning horizon is large, or infinite.

5.2. Some Related Literature

The *essential elements* of a dynamic optimisation model for investment with growth (Islam, 2001a):

i) the length of the planning horizon,

ii) an optimization criterion,
iii) the form of intertemporal time preference or discounting,
iv) the structure of dynamic systems under modelling,
iv) the initial and terminal conditions.

There is extensive literature on methodology and practice for economic areas such as optimal growth (see e.g. Islam, 2001a; Chakravarty, 1979), and on investment models (see e.g. Campbell and Viceira, 2002 for a recent account, also Karatzas, 1996 for modelling in continuous time, and dynamic optimisation modelling in Taperio, 1998; Sengupta and Fanchon, 1997; Vickson and Ziemba, 1975)), some questions remain concerning the mutual interaction of these areas.

Dutta (1993) provides an up to date survey of results concerning the elements listed. The element (i) concerns the length of the planning horizon, the associated choice of a final period capital stock, and the possible choice of an infinite horizon. Dutta (1993) cites some results of Brock (1997), Mitra (1983), and Mitra and Ray (1984), that establish, under some assumptions including the continuity, convexity, etc. mentioned above, the insensitivity to horizon length, and the existence of a finite-horizon plan close to a given infinite-horizon plan. But this closeness does not always hold; there are counter-examples. Significant questions are whether a finite-horizon investment plan is relatively insensitive to the exact choice of horizon length, and whether each finite-horizon optimal plan is close to some infinite-horizon optimum?

In (ii), optimization requires some *objective function* to be maximized (or minimized). Often, a utility function of consumption alone is chosen, usually satisfying properties of continuity, monotonicity, and concavity. However, it may be appropriate to consider several objectives (see discussion in section 4b), and to relax the concavity requirement. Other optimality criteria, not often used for financial modelling include overtaking criterion, catching-up criterion, maximum criterion, etc. (Leonard and Long, 1992); there is some discussion in section 5.4.

For (iii), an infinite horizon generally requires some discount factor, but not necessarily the usual exponential factor, since this assigns negligible value to the far future. Alternative discount factors (see Heal 1998, Islam and Craven, 2001b) may be considered, For (iv), terminal conditions are often given by transversality conditions the value of the terminal year capital stock is specified, perhaps as zero (see Leonard and Long, 1998 for a survey).

The next paragraphs outline a number of specific financial models, discussed in the literature. A recent discussion of computation of optimal portfolio choice problems may be seen in Islam and Oh (2003), and Oh and Islam (2001).

In more general financial models (see Sengupta and Fanchon 1997), the growth of wealth w_t at times $t = 0, 1, 2, \ldots$. is described by a dynamic equation $w_{t+1} = r_t(w_t - c_t)$, in which c_t is consumption at time t, and r_t is a stochastic term, perhaps described by a first order Markov process. They use a logarithmic utility function, and maximize the present value: $\sum \rho^t \log(u_t)$.

In the model of Hakansson (1975), there are M *investment opportunities* $i = 1, 2, \ldots, M$. A dynamic equation, in discrete times $t = 1, 2, \ldots$, relates capital x_t at the start of period t to the amount invested by:

$$x_{t+1} = \sum_i (\beta_i - 1 - \rho) z_{it} + (1 + \rho)(x_t - c_t) + y_t \ ,$$

where ρ is interest rate (Hakansson's quantity $r = -1 + \rho$); β_i is the stochastic factor which multiplies capital for investment i; y_t = income from non-capital sources; c_t represents consumption; some assumptions are made on the β_i . An objective function:

$$\sum_t \alpha^t u(\sum_i ((\beta_i - 1 - \rho) z_{it} (1 + \rho)(x_t - c_t) + y_t = \sum_t \alpha^t u(x_{t+1})$$

is to be maximized, presumably with respect to variables z_{ij} . But it is not clear what bounds are placed on the z_{ij} ; and the costs of investment capital are not explicit. Hakansson obtained an analytic optimum using dynamic programming.

The asset pricing model of Lucas (1978) was summarized in Judd (1998). In this model, a single asset pays dividends according to the ($AR(1)$) stochastic model $y_{t+1} = r y_t + e_{t+1}$, with independent identically distributed random variables e_t . The present value of utilities $u(c_t)$ is maximized. The equilibrium prices p_t satisfy:

$$u'(y_t) p_t = b\mathbf{E}\{u'(y_{t+1})(y_{t+1} + p_{t+1|} p_t\},$$

where b is a discount factor. This leads to an integral equation for the prices p_t . The dividends y_t perhaps represent the consumption c_t .

In another version of the model (see Gourieroux and Janiak, 2001), the present value of utility functions of consumption is maximized, subject to a "budget constraint":

$$q_t C_t + a_t p_t = R_t + a_{t+1} p_t,$$

where q_t denotes price of the (single) consumption good, a_t denotes a vector of allocation of investment, C_t denotes consumption, R_t denotes external income, p_t is (presumably) the market price (vector) of the investments at time t. An unconstrained optimum was sought, by setting the gradient to zero.

Malliaris and Brock (1982) describe Merton's model, in which the share prices are described by a stochastic process in continuous time. Consumption and amount of new investment are decided by current prices. The objective to be optimized is an integral of utility of consumption over the planning period, plus a function of wealth at the end time. All income is supposed to come from capital gains. A solution is obtained by stochastic dynamic programming. Malliaris and Brock also discuss an *asset pricing model* , based on Lucas

(1978), which assumes the *hypothesis of rational expectations* (a hypothesis on investors' behaviour), and involves stochastic terms. The existing dynamic financial models (including the ones discussed above) are not specified with an explicit discussion of a set of the elements of a dynamic economic model. The above set of elements are chosen on an ad hoc basis in the existing literature.

5.3. Some Approaches

For such stochastic models in continuous time, dynamic programming is a useful theoretical technique, but it presents difficulties for numerical computation. From a methodological viewpoint, a computation would require discretization of both time and levels of prices, and arguably one might as well start with a discrete model. However, it may be questioned whether investors, or markets, indeed behave rationally according to Lucas's model, or any other. Moreover, the growth factors, such as Hakansson's β_i , are extrinsic to the models, and are thus not analysed. But growth rate should depend on physical capital, as discussed below in section 5.4, and investment, and so may be more specifically modeled. A distinction is necessary between physical capital and its current (and fluctuating) market value.

These dynamic models have an inherent multiobjective character. One may wish to maximize both consumption and capital growth, both for the near future and the more distant future. These objectives are in conflict, and an improvement on one may not compensate for a deterioration in another. A model may include several parameters, specifying the relative weight given to the conflicting factors, Thus, for example, a model could maximize the capital at a horizon time, plus a proportion σ_1 of an estimated future growth, subject to a floor on consumption, thus consumption $\geq \sigma_2$ at all times. The result of varying the parameters σ_1 and σ_2 would then need study, usually computationally.

Suppose there are objective functions $f_i(z)$ $(i = 1, 2, \ldots)$ to be maximized, subject to constraints $g_j(z) \geq 0$ $(j = 1, 2, \ldots, m)$. Some of the f_j could be *end-point terms*, describing the far-future. As well as a Pareto maximum problem for the vector objective $f = (f_1, f_2, \ldots .)$, other versions of the problem would turn one or more objectives into constraints, such as $f_j(z) \geq \sigma_j$. (For the present, the parameter σ_j is absorbed into f_j , so $f_j(z) \geq 0$ is written). For all of these related problems, the Lagrangian:

$$L = \sum \tau_i f_i(z) + \sum \lambda_j g_j \text{ (z)}$$

is the same, and the Karush-Kuhn-Tucker (KKT) necessary conditions for an optimum at $z = p$ are the same, *except* that the multipliers τ_i and λ_j may take different values. If hypotheses are assumed that make the KKT conditions also sufficient for an optimum, then all of these problems are optimized at different Pareto points of the Pareto maximum problem for f. The hypotheses could be that all the $-f_i$ and $-g_j$ are *invex* with respect to the same scale function, or that all the f_i and g_j are *pseudoconcave* . Equivalently, the vector function Q, whose components are all the $-f_i$ and $-g_j$, is *invex* if, for all x and p:

$$Q(x) - Q(p) \geq Q'(p)\eta(x,p)$$

for some scale function η. Convexity is the case where $\eta(x,p) = x - p$. However, *invexity* is enough to ensure that the necessary KKT conditions are also sufficient for an optimum (see e.g. Craven, 1998).

However, *invex* may be difficult to verify for an optimal control problem, because the dynamic equation, an equality constraint, presents difficulties (see Craven, 1998).

5.4.1. A proposed model for interqction between financial investment and physical capital

An increase in share price does not directly increase productivity. Indirectly (and after some time delay) it may do so, by allowing additional capital to be raised by a share or bond issue. The present model considers productivity to depend on the physical capital (which is unchanged by a fluctuation in share price), plus additional investment, which may be explicitly modelled. The latter are of two general kinds, namely *internal investment* (that part of the profits which is used to increase physical capital, including raw materials, etc.), and *external investment* (from share issues, etc.)

How does physical capital affect the share price, and how does the share price affect *external investment* . If these aspects are to be modelled (instead of putting some unexamined *growth factor* into the model), then some assumptions must be made. Some simple assumptions are the following. The share price could be assumed to follow a random walk, say jumping up or down by an amount d with probability q for each jump, with d and q adjusted to match an observed variance of share price. But the share price should tend to increase somewhat with an increase (or positive rate of increase) of the physical capital. Perhaps the probability of jumping up by d should depend on the physical capital? The *external investment* , it could be assumed to increase as share price increases (or perhaps when its rate of increase gets over a threshold?) Note that this is a positive feedback path, hence tending to instability, and unlikely to follow a linear law. These actual phenomena seem to fall outside the usual market model assumptions, but they may still be modelled.

Consider investment opportunities $j = 0, 1, \ldots, n$, where $j = 0$ is "risk-free" (i.e. a bank account, whose balance may be positive or negative). At time $t = 0, 1, 2, \ldots, T$, let k_{tj} = physical capital (measured by historical investment, in constant dollars); let m_{tj} =market price; so "wealth" consists of terms $m_{tj}k_{tj}$. Instead of assuming an exogenous growth term, growth will be modelled, by:

$$k_{t+1,j} = (1 - d)[k_{tj} + g(bk_{tj} - c_{tj} + y_{tj}],$$

where c_{tj} = consumption (or sales, exports, etc.), y_{tj} denotes new investment (e.g. from a share issue), g is an increasing concave function, describing a *law of diminishing returns,* d is a depreciation factor, and $b(< 1)$ describes the fraction of capital that is available for growth. (This model is crude, since physical capital is of several kinds.)

The *wealth* at time t consists of terms $m_{tj}k_{tj}$. Fluctuations in the market prices m_{tj} do not directly change the physical capital, on which growth depends, but they will affect the investment term y_{tj} (which requires modelling in terms of the increase in *wealth* at a somewhat earlier time). The function g is analogous to the concave utility functions used by other authors, but is here applied to the capital available for growth. The bank account ($j = 0$) allows for additional investment (subject to paying interest), or for selling assets without reinvesting them. If z_{tj} units of investment j are held at time t, and v_{tj} are purchased during the period $(t, t + 1)$ (with $v_{tj} < 0$ for a sale), then:

$$v_{t0} + m_{t1}v_{t1} + \ldots + m_{tn}v_{tn} = 0 \;;$$

$$z_{t+1,j} = z_{tj} + v_{tj} \quad (j = 1, 2, \ldots, n);$$

$$z_{t+1,0} = [z_{t0} + m_{t1}q_{t1} + \ldots + m_{tn}q_{tn} \;]$$

Here q_{tj} describes any dividend or interest, or repayment of a bond.

Consider first a simple model for market prices:

$$m_{t+1,i} = m_{t,i} + e_{t,i} + \zeta[k_{t,i} - k_{t-1,i}],$$

in which the $m_{t,i}(t = 0, 1, 2, \ldots)$ are independent identically distributed random variables with a given variance σ^2, and ζ is a small positive constant. Thus, the price is modelled as a random walk, with an additional small *feedback term* depending on the change of physical capital. This feedback may produce instability – but note that market prices are, in fact, often unstable. The different investments j are considered here as independent, though this assumption may need modification.

The new investment term y_{tj} may be modelled as:

$$y_{tj} = \zeta'[m_{tj}k_{tj} - m_{t-1,j}k_{t-1,j}]_+ \;,$$

where ζ' is a small parameter, and $[w]_+ := w$ if $w \geq 0$, 0 if $w < 0$. Note that the time lag, here one interval, is another parameter that may be changed.

As well as this dynamic model, appropriate objective function and constraints must be specified. One specification is to maximize the expectation of the wealth at the horizon time T :

$$\mathbf{E} \sum_j m_{Tj}k_{Tj},$$

subject to a constraint on consumption:

$$c_{tj} \geq b_j + b'_j t \quad (t = 0, 1, \ldots, T; j = 1, 2, \ldots \;, \mathrm{n}) \;.$$

There might also be a cost term $\sum_{t,j} \mu[c^* - c_{tj}]$ subtracted from the objective, to describe a penalty cost imposed when the consumption is allowed to fall below some target level c^*. This formulation assumes that the *far future* (after time T) is accounted for by maximizing the wealth at time T. For each investment j, there is a floor under consumption, which increases with time, at a rate given by the parameters b'_j .

Another specification maximizes a consumption objective:

$$\sum_j r^t U(c_{tj}) + \sum_j \sigma_1 j U(c_{jT}) r^T/(1-r) ,$$

in which $U(.)$ is a (concave increasing) utility function, subject to the above floor on consumption, and to a terminal constraint on wealth:

$$\mathbf{E} \sum_j m_{Tj} k_{Tj} \geq \sigma_3.$$

Here the parameters σ_{1j} relate to the balance between short term and far-future, and σ_3 relates to the balance between consumption and capital.

For the discrete-time optimal-control financial models considered here, the existence of an optimum is automatic, since the feasible region is bounded, and continuous functions are involved. This existence result does not hold for a continuous-time optimal control financial model, since infinite dimensions are involved. However, if the necessary Pontryagin conditions are solvable (as is the case for all the continuous-time models considered in this chapter), then that solution is the optimum if either the solution is unique, or if a generalized-convexity condition such as *invexity* (see section 3b) holds as in many models of finance (and also economic growth), but uniqueness does not generally hold.

5.5. A computed model with small stochastic term

In this section, a financial model is developed and implemented, incorporating explicit relations between physical and financial capital, and so allows the extent of dependence of these two to be analysed.

For discrete time $t = 0, 1, 2, \ldots$, denote k_t = physical capital, c_t = consumption, q_t = external investment, c^* = consumption target, m_t = market price for shares, ρ = depreciation rate, and $\varphi(s) = a(1 - e^{-bs})$ is a concave increasing *growth function.* Consider the following dynamic equations for physical capital (which may include knowledge and human capital) and market price:

$$\Delta k_t = -\rho k_t + \varphi(k_t - c_t + q_t), \ k_0 \text{ given}$$

$$\Delta m_t = \theta m_t + \mu \Delta k_t + \xi_t, \ m_0 \text{ given.}$$

Here the coefficients θ, and $\mu > 0$, describe a market price trend and an influence of observed physical capital on market price; the ξ_t are i.i.d. random variables with mean zero. (A more complicated, Markovian, model could also be considered.) In particular, if the random terms are negligible and if $\theta = 0$, then:

$$m_t = m_0 + \mu(k_t - k_0).$$

The interaction between financial capital, namely the market value $m_t k_t$, and external investment q_t is assumed to have the form $q_t = \epsilon[\Delta(m_{t-1}k_{t-1})]_+$. Here ϵ is a small positive coefficient, and $[.]_+$ makes only increases in market value have an effect. There is a time lag, here of one unit. (This model does not discuss speculative investment.) Denote the expectation $\bar{k}_t := \mathbf{E}k_t$, also $y_t := k_t - \bar{k}_t$, and $\bar{m}_t := \mathbf{E}m_t$.

Substitution of q_t into the equation for Δk_t gives:

$$k_{t+1} = (1-\rho)k_t + \varphi(k_t - c_t + \epsilon[\Delta(m_{t-1}k_{t-1})]_+).$$

The contribution of y_t to the right side of this equation contains linear terms, whose expectation is 0, and a quadratic term, approximated by $\frac{1}{2}\varphi''(\bar{k}_t - c_t)\theta_3 \bar{m}_t y_t^2$. Hence, denoting $\nu_t := sgn\Delta(\bar{m}_{t-1}\bar{k}_{t-1})$,

$$\bar{k}_{t+1} \approx (1-\rho)\bar{k}_t + \varphi(\bar{k}_t - c_t + \epsilon[\Delta(\bar{m}_{t-1}\bar{k}_{t-1})]_+)$$

$$+\frac{1}{2}\epsilon\bar{\varphi}''(\bar{k}_t - c_t)\bar{m}_t\nu_t \text{ var } k_t,$$

Approximating the contribution to y_{t+1} from ξ_t by the linear term

$$(1-\rho)\xi_t + \varphi'(\bar{k}_t - c_t)(\xi_t + \epsilon\bar{k}_t\xi_{t-1})$$

gives:

$$\text{var}k_{t+1} \approx (1-\rho+\varphi'(\bar{k}_t - c_t)) \text{ var } k_t + \nu_t[\epsilon\varphi'(\bar{k}_t - c_t)\bar{k}_t]^2 \text{ var } m_t.$$

These approximations may be appropriate when ϵ is small.

The interactions described by μ (effect of physical capital on market price) and ϵ (effect of market price on external investment) represent a *feedback* process, whereby an increase of physical capital tends to produce a further such increase.

The objective function to be maximized is taken as:

$$\bar{k}_{T+1} - \omega\sum_{t=0}^{T}[c^* - c_t]^2,$$

The elements of this model are (i) the finite planning horizon T, (ii) the objective function, a utility function involving terminal capital and consumption, with a penalty when consumption falls too low (ω is a weighting factor), and (iii) boundary conditions, as upper and lower bounds on consumption.

For a computed example, consider the numerical values:

$$T = 20, \rho = 0.1, a = 4.0, b = 0.1, c = 0.4, \omega = 0.5, \epsilon = 3.0, \theta = 0.$$

The cases with $s :=$ var $m_t = 0.0,\ 0.3$, and 0.6 were computed, each with the parameter μ as either zero (no influence of physical capital on market price, or $\mu = 0.02$ (physical capital having a small positive influence on market price). The parameters for the several computations are tabulated as follows (ıthe captions to the figures refer to the computer run numbers):

Parameters for Figures 1 to 6

$\kappa = 100\mu$, where μ is the feedback factor; $\omega = 0.5$ throughout.

Run	1	2	3	4	5	6	7	8	9	10	11	12	13	14
s	.0	.3	.6	.0	.3	.6	.3	.3	.3	.3	.6	.6	.6	.6
κ	2	2	2	0	0	0	2	5	10	50	2	5	10	50

Figure 1 shows the growth of expectation of physical capital with time, with the different levels of variance s distinguished by solid, dashed, and dotted lines, and the plotted points with $\theta_2 > 0$ marked by + signs. Also shown are the standard deviation $\sqrt{(}\text{var } k_t)$ of physical capital. Figure 2 shows the consumption, with the same convention of plotting. Figure 3 shows expectations of financial capital and physical capital, for the case $s = 0.3$ and $\theta_2 = 0.02$.

In Figure 1, an increase of the variance of the market price increases the physical capital, and decreases its standard deviation. The *positive feedback,* when μ is positive, increases the physical capital (Figure 1), and also the financial capital (Figure 3), as well as the consumption (Figure 2). For these assumed parameters, the expected market price $\bar{m}_t = m_0 + \mu(\bar{k}_t - k_0)$ increases by 40% over the time period considered. Note also also that financial capital $m_t k_t$ has more random variation than physical capital k_t .

Figures 4,5 and 6 show in more detail the results of the feedback factor μ (with ϵ constant at 0.3) and the variance of the price m_t . As μ increases, the expectation of physical capital increases (Figure 4), the consumption increases (Figure 5), but the variability of physical capital, as measured by its Standard Deviation, decreases. In Figures 4 and 5, the curves for $s = 0.3$ and $s = 0.6$ have been separated by moving the latter curves down.

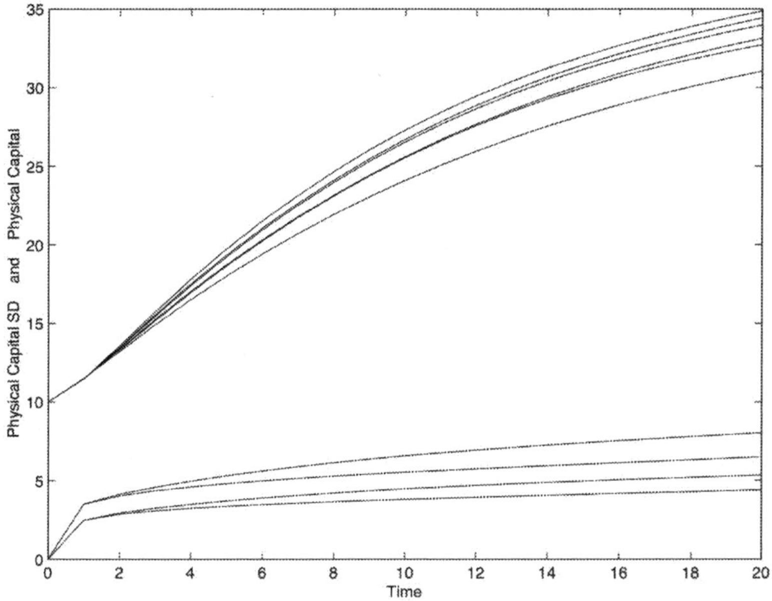

Fig. 1 - Physical Capital and its Standard Deviation
From top: Capital 1,4,2,5,3,6.
Standard deviation 6,3,5,2 ; 1 and 4 are zero.

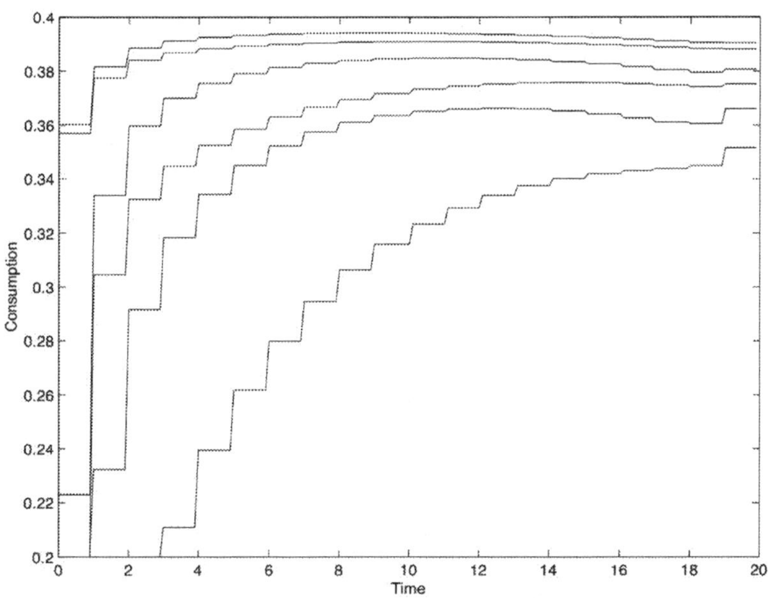

Fig. 2 - Consumption
From top: Consumption 4,1,2,5,3,6.

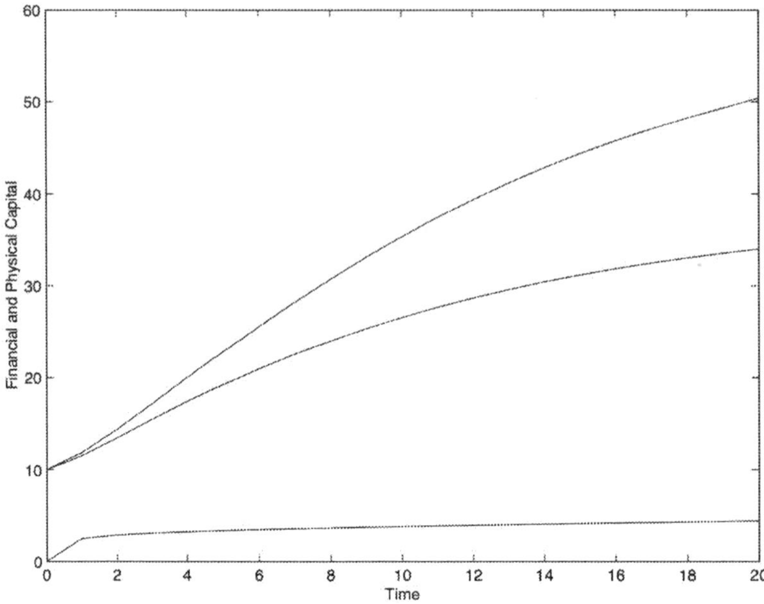

Fig. 3 - Financial and Physical Capital
The graph of Financial Capital lies above that of Physical Capital.

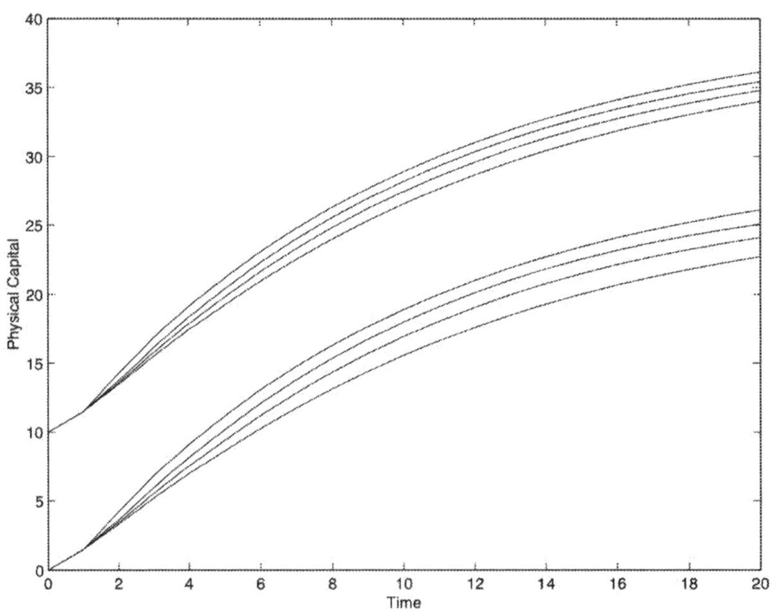

Fig. 4 - Effect of Feedback on Physical Capital
From top: Physical Capital 10,9,8,7, then 14,13,12,11.

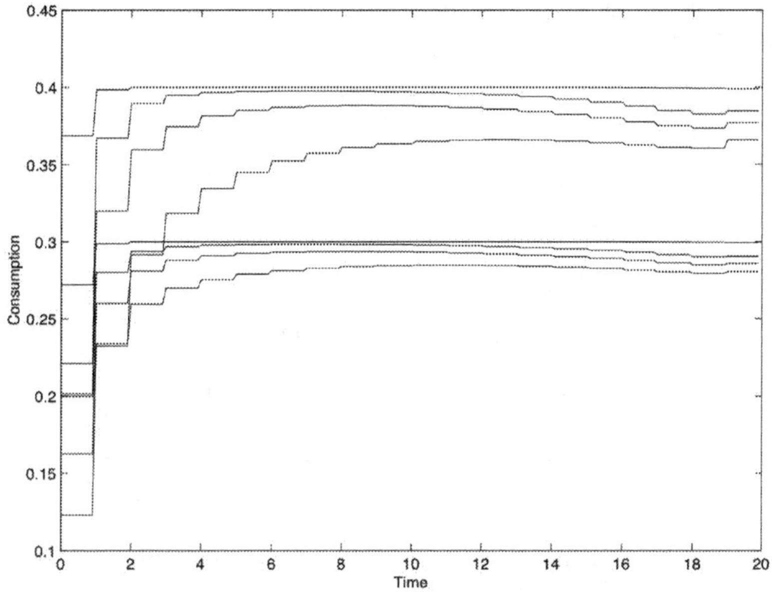

Fig. 5 - Effect of Feedback on Consumption

From top: Consumption 14, 13, 12, 11, then 10, 9, 8, 7 with 0.1 subtracted.

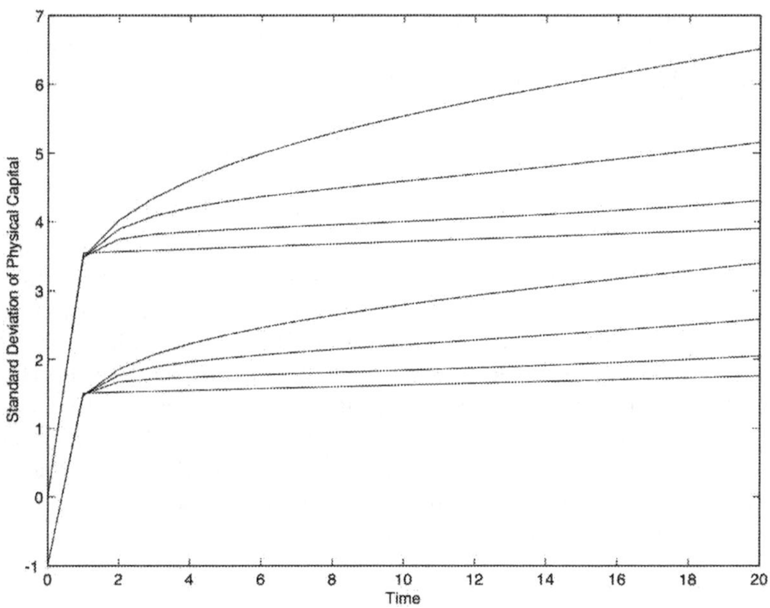

Fig. 6 - Effect of Feedback on Variability of Capital

From top: Standard Deviations 11, 12, 13, 14,
then 7, 8, 9, 10 with 1.0 subtracted.

The final capital k_T increases with increase of the initial slope $\varphi'(0)$ of the growth function, as shown by the sample figures (computed with $\epsilon = 0.1$ and zero stochastic term):

ρ	k_0	a	b	ab	c_0	c_T	k_{T+1}
0.1	10	4.0	0.1	0.4	.941	.985	33.49
0.1	10	4.0	0.2	0.8	.975	.999	36.09
0.2	10	4.0	0.1	0.4	.970	.879	14.62
0.1	10	3.0	0.01	0.03	.293	.271	2.21
0.1	5	3.0	0.01	0.03	.293	.270	1.07

Numerical values for parameters were chosen to exhibit the possible behaviours. No empirical data is available for some of the parameters, especially μ and ϵ.

These results demonstrate the interdependence of physical and financial capital, in terms of accumulation and dynamics. It is observed that increased uncertainty reduces the level of consumption.

The feedback interactions lead to higher levels of both capital and consumption, and also tend to smooth consumption profiles over time. In some circumstances the consumption is reduced to its lower bound, for part of the planning period, in order to achieve the target for terminal capital.

5.6. Multiple steady states in a dynamic financial model[8]

An optimal control model need not be very complicated before (unlike the simpler economic models) there may be more than one stationary solution. The following is a mathematical example to illustrate this. A continuous-time (deterministic) model is given here, to simplify formulas. Results would be qualitatively similar for a discrete-time model, obtained by discretizing the time interval $[0, T]$ into suitable subintervals. In the example, the state $x(t)$ might describe capital stock, and the control $u(t)$ might describe the fraction of capital stock to be given for consumption. If some parameter of the model were varied (presumably at a slower rate than the model tends to a steady state), then this might result in a jump from one solution to another. There is a possibility here to model discontinuities in the real world.

Consider an optimal control financial model in continuous time:

$$\text{MIN} \int_0^T e^{-\rho t} f(x(t), u(t)dt \text{ subject to } x(0) = x_0, \dot{x}(t) = m(x(t), u(t)).$$

8 See also section 2.5.

As with various economic growth models, no active constraints are assumed for the control $u(t)$. The time horizon T is assumed to be fairly large. But the discount factor $e^{-\rho t}$ prevents a steady state being reached. The Hamiltonian is $e^{-\rho t} f(x(t), u(t)) + \lambda(t) m(x(t), u(t)$. In terms of the *steady state Hamiltonian* $f(x(t), u(t)) + \Lambda(t) m(x(t), u(t))$, where $\Lambda(t) = e^{\rho t} \lambda(t)$, the adjoint differential equation becomes:

$$-\dot{\Lambda}(t) = f_x(x(t), u(t)) + \Lambda(t) m_x(x(t), u(t)) - \rho \Lambda(t)$$

(with no boundary condition), and Pontryagin's principle (in the absence of active control constraints) requires that $f_u(x(t), u(t)) + \Lambda(t) m_u(x(t), u(t)) = 0$. A *steady state,* in which $x(t)$, u(t) and $\Lambda(t)$ take constant values, may be sought by equating $\dot{x}(t)$ and $\dot{\Lambda}(t)$ to 0.

Can fairly simple functions f and m give rise to multiple steady states? Consider, in particular, $m(x, u) = \Phi(x) -$ ux, where Φ is a cubic, say $\Phi(x) = a + (3/2)x^2 - x^3$ (assuming some scaling for x). Then $\Phi(x) - ux = 0$ when a $-ux + (3/2)x^2 - x^3 = 0$, and this cubic will have three real roots $x = x(u)$, for an interval of values of u, In particular, if $a = 0$, then:

$$x(u) = 0 \text{ or } (1/2)(3 \pm \sqrt{(9 - 16u)}), \text{ with } u < 9/16.$$

Consider, in particular,

$$f(x, u) = (1/2)(x^2 + u^2).$$

Then the *steady state* (if it exists) requires that:

$$\Phi(x) - ux = 0; x + \Lambda(\Phi'(x) - u - \rho) = 0; u + \Lambda(-x) = 0.$$

Hence x depends on u as discussed above, $\Lambda = u/x$, and

$$0 = x + (u/x)(3x - 3x^2 - u - \rho) = 0.$$

Hence x must satisfy the equation:

$$0 = x^2 + (\Phi(x)/x)(3x - 3x^2 - \rho) - \Phi(x)^2/x.$$

If, in particular, $a = 0$, and $\rho = 0$ (no discounting), then x satisfies a cubic equation, so that $x = 0$ or x satisfies:

$$1 + (3/2 - \text{x})(3 - 3\text{x}) - x(3/2 - x)^2 = 0.$$

Another instance of multiple steady states is given in Islam and Craven (2001).

5.7. Sensitivity questions, concerning infinite horizons

Consider first a single-objective financial optimization problem:

$$\mathrm{MAX}_z F(z,q)) \text{ subject to } g(z,q) \leq 0,\ \ k(z,q) = 0,$$

where q is a perturbation parameter, and suppose that, when $q = 0$, a *strict maximum* is reached at $z = \bar{z}(0)$. This means that, for small enough $r > 0$:

$$[g(z,0) \leq\ 0,\ k(z,0) = 0, \|q\| = r\ ]\ \Rightarrow F(z) \leq F(\bar{z}(0)) - \delta(r)$$

for some $\delta(r)\ >\ 0$. Given also a continuity condition,, the *strict maximum* implies that, for sufficiently small $\|q\|$, the perturbed problem reaches a local maximum at $\bar{z}(q)$ close to $\bar{z}(0)$, thus $\bar{z}(q) \rightarrow z(0)$ as $q \rightarrow 0$. Clearly, this does not usually happen for a maximum that is not strict,

The additional conditions required for this result (see Craven, 1994) are that (F, g, k) is uniformly continuous in a local region around the point $(z, q) = (\bar{z}(0), 0)$, there are feasible solutions to the perturbed constraints, and, when $q \neq 0$, $F(., q)$ reaches a minimum on each closed bounded set (this last would usually be assumed on economic grounds).

For an optimal control problem with state $x(t)$ and control $u(t)$, the dynamic equation:

$$x(0) = x_0\ ,\ \dot{x}(t) = m(x(t), u(t), t)\ (\mathrm{t}\ \geq 0)$$

is assumed to determine $x(.)$ as a function of $u(.)$. So an objective function:

$$\int_0^T f(x(t), u(t), t)dt + \Phi(x(T))$$

may be expressed as $J(u(.))$, to be maximized subject to constraints on $u(.)$. The horizon T will be written as 1/q, where the perturbation parameter $q = 0$ for an infinite horizon.

The uniform continuity condition is then not automatic. It is fulfilled (see section 5.10) if f is quadratic, and high-frequency oscillations of the control $u(t)$ are suitably small.

Assume now that an optimum is reached at $z(t)\ \equiv\ (x(t), u(t)\ =\ p(t)$. Several cases arise for the infinite horizon.

Case $J(p)$ finite. This is only likely if f includes a discount term. If $z(t)$ tends to a limit, q say, as t tends to infinity, $f(q)$ will sometimes (depending on the rate of convergence) be the dominating factor in $J(p)$. Under some (substantial) boundedness conditions, a suitable time transformation $t = k(s)$ (with $k(.)$ increasing and $k(0) = 0$) maps the infinite time interval to [0,1], but may produce an unbounded integrand. If $f(z(t), t)$ tends fast enough to zero as $t \rightarrow \infty$, an equivalent control problem is obtained, over a finite interval of transformed time s, and the Pontryagin theory applies to it. If the maximum for the infinite horizon is strict, and the uniform continuity condition holds, then the optimum objective for horizon T will be close to the optimal objective for infinite horizon, provided that T is sufficiently large.

Periodic case. For some finite T, impose the boundary condition that $z(T) = z(0)$. Assume no discount factor. Then the problem may be optimized over [0, T], then it effectively starts again at $t = T$ for a new interval $[T, 2T]$, and so on.

Case $J(p)$ infinite, other than the periodic case. Since maximizing $J(z)$ is now meaningless, suppose that $T^{-1} \int_0^T f(z(t), t)dt \to b$ as $T \to \infty$, and then seek to maximize $b(z)$. Can one maximize $Q(T) := T^{-1} \int_0^T f(z(t), t)dt$ for some large enough fixed T, and derive from this some approximation to the required maximum b? For this purpose, assume that the given problem reaches a *strict maximum* of b^* at $z = p$. Assuming also the uniform continuity condition, then $1/T$ may be regarded as a perturbation parameter, applied to $Q(T)$ instead of $J(T)$, leading to a maximum of $Q(T)$ at some point $p(T)$ close to p^*.

Catching up criterion. Under this criterion, $p(t) = (\bar{x}(t), \bar{u}(t))$ is optimum if:

$$\mathrm{limsup}_{T \to \infty} \int_0^T [f(z(t), t) - f(p(t), t)]dt \leq 0,$$

for feasible plans $(x(t), u(t))$. Equivalently, $T^{-1} \int_0^T f(z(t), t)dt$ may be discussed, though a limit as $T \to \infty$ is not necessarily assumed.

However, the question of how large T must be, in order to closely approximate the infinite horizon case, calls for further study.

While these sensitivity results are stated for continuous-time optimal control models, the ideas also apply to discrete-time control models, such as the dynamic model of section 5.5 with a consumption objective.

5.8. Some conclusions

Dynamic optimization models in finance can be usefully extended in several directions, leading to a more integrated and operational approach. The aspects considered include a systematic analysis of the elements of the model, the role of physical capital, as distinct from fluctuating market value, some attempts to model investment, sensitivity to time horizons, possible multiple steady states, and some numerical calculations. A dynamic optimization model requires a specification of a standard set of elements for a growth model.

While an infinite time horizon may not be thought relevant to financial modelling, a long horizon may be, when a continuing enterprise is considered. Computation of a simple model, showing the interaction between physical capital, financial capital, and investment, have led to several qualitative conclusions, detailed above.

Some issues remaining are to relate the model to empirical data, and to explore the relevance of mathematical concepts such as *generalized convexity* (particularly *invexity*) to these financial models.

A financial model of a different kind is the model by Davis and Elzinga

(1972) for dividends and investment in a utility company. A computation of this model using the SCOM package is given in section 3.7; a stochastic version is considered in Craven (2002).

The usefulness of MATLAB packages for such computations is not speed, but rather a great facility in changing model details and parameters, without needing extensive programming. The step-function approximations used here have been shown to be adequate, both theoretically (Craven, 1995) and computationally (Teo, Goh and Wang, 1990).

5.9. The MATLAB codes

These are the MATLAB codes used for the discrete-time optimal control mode, including a small stochastic contribution. For purpose of computation, the difference equations have been approximated by differential equations, enabling continuous-time optimal control software to be used.

The time scale: $t = 0, 1, 2, \ldots, n$ translates to $t = 1, 2, \ldots, nn + 1$ in MATLAB M-code. The MATLAB package *constr* is called to optimize, given the functions from the following subroutines. The endpoint condition on the state (capital) is computed as a penalty term. The parameters ω and μ in the model appear in these codes as mu and fb respectively.

```
% fqs_run.m   % Calling program
subs=cell(1,9);
subs={'fqsx','fqj','fqf','fqc'};
par=[4, 1, 20, 0,0 ,20, 0.1, 3.0, 4.0, 0.1, 0.4, 0.5, 0.6, 1, 0.02];
% nx  nu  nn  0  0  T  rho eps  a  b  c  omega  s  switch  mu
% nx = n umber of states, nu = number of controls, nn = number of
% intervals
nn=par(3); nu=par(2); nx=par(1);
xinit=[10.0, 10.0, 0.0, 0.0];   % Initial conditions for states
u0=0.2*ones(nn,nu); ul=0.2*ones(nn,nu);
%ul=zeros(nn,nu);
uu=10*ones(nn,nu);
Control=constr('fqd',u0,[],ul,uu,[],par,subs,xinit)
[jf,g,state]=feval('fqd',Control,par,subs,xinit);
state
jf
y1=Control; y2=state(:,1); y3=sqrt(state(:,4));

function [f,g,state] = fqd(um, par, subs, xinit)   % Organize computation
nx=par(1); nu=par(2); nn=par(3);b=par(6);
ıpd=char(subs(1));
xm=zeros(nn+1,nx); xm(1,3́a)=xinit;
for tc=1 : nn
ff=feval(pd,tc,um,xm,par);
xm(tc+1,3́a) = xm(tc,:) + ff;
```

```
end
% compute objective function
jf=0; pd=char(subs(2));
for tc=1 : nn
ff=feval(pd,tc,um,xm,par);
jf=jf+ff;
end
pd=char(subs(3));
jf=jf+feval(pd,tc,um,xm,par);
pd=char(subs(4));
for tc=13ann
g(tc)=feval(pd,tc,um,xm,par);
end
f=jf;
state=xm;

function ff=fqsx(t,um,xm,par)  % RHS of dynamic equations
nn=par(3);T=par(6); hh=T/nn; cc=par(11);mu=par(12); fb=par(15);
del=par(7);eps=par(8);aa=par(9);bb=par(10);sig2=par(13);
mt=1.0+fb*(xm(t,1)-xm(1,1));mm=mt+fb*xm(t,2);
temp=xm(t,1)-um(t,1)+eps*mm*max(xm(t,1)-xm(t,2),0);
swit=par(14);
ph1=aa*bb*exp(-bb*temp);
ph2=-bb*ph1;
temp2=eps*ph1*xm(t,1);
ff(1)=(-del)*xm(t,1)-aa*(exp(-bb*temp)-1)
                    +swit*eps*mt*ph2*xm(t,4);  % Capital
ff(2)=xm(t,1)-xm(t,2);  % Relates to differences of Capital
ff(3)=max(cc-um(t,1),0)^2; cc-um(t,1);
ff(4)=(-del+ph1)*(-del+ph1)*xm(t,4)+swit*temp2* temp2*sig2;

function ff=fqj(t,um,xm,par)  % Objective integrand
nn=par(3);T=par(6); hh=T/nn; cc=par(11);
del=par(7);eps=par(8);aa=par(9);bb=par(10);
ff=0;

function ff=fqf(t,um,xm,par)  % Objective endpoint term
nn=par(3);T=par(6); hh=T/nn; cc=par(11);
del=par(7);eps=par(8);aa=par(9);bb=par(10);
mu=par(12);
ff=-xm(nn+1,1)+mu*xm(nn+1,3);
xm(nn+1,3);
```

5.10. Continuity requirement for stability

For a control problem on an infinite time domain, assume that the control $u(t)$ satisfies $\|u\|_2 = [\int_0^\infty |u(t)|^2 dt < \infty$. Denote by $Qu(s)$ the Fourier transform of

$u(.)$. Now assume, more restrictively, that $\|b(.)Qu(.)\|_2$ is finite, for a region around the optimal $u(.)$, where the weighting function $b(s)$ is positive, and $b(s) \to \infty$ as $s \to \infty$. This places a serious restriction on high-frequency oscillations of $u(t)$. Assuming this, the *uniform continuity* requirement for the sensitivity result follows for functions f which are linear or quadratic (see Craven, 1994), and hence for some other functions whose growth rate is no more rapid than these.

Chapter 6

Modelling Sustainable Development

6.1. Introduction

The issues of *sustainability of growth,* and *social welfare* that ensures intergenerational equity, are controversial, and form an important area of study in social welfare and social choice in contemporary models. In recent literature, Hamiltonian-based measures of social welfare have been used to define the concept of sustainable growth In the mainstream economic interpretation, where social welfare is measured in terms of pure economic variables such as income or consumption, *sustainable growth* relates to economic conditions of non-declining consumption or capital (whether man-made, natural, or environmental) over time. *Sustainable growth* has been formalised in different ways. In the optimal growth literature, sustainable consumption is characterized by the *golden age*, representing the maximum consumption possible without reducing the potential for the same level of consumption in the future. Given the numerical orientation of this study, the objective function here has cardinality, measurability and intergenerational utility and welfare comparability implications. Several other definitions of sustainability have been given (Faucheux, Pearce and Proops, 1996). Conceptual and theoretical studies of sustainability are well advanced. In spite of several operational numerical studies of this issue (e.g. Islam, 2001a), there is still a strong need for operational methods for sustainability modelling. This chapter presents several operational models and methods for sustainable growth and welfare, in order to explore a range of optimal control models in welfare economics.

6.2. Welfare measures and models for sustainability[9]

Various theoretical articles (Heal, 1998; Faucheux, Pearce and Proops, 1996; Smulders, 1994) have incorporated sustainability in optimal growth models. In terms of optimal control theory, sustainable consumption or welfare for an autonomous infinite-horizon problem have been described by a Hamiltonian function (see Leonard and Long, 1992; and Heal, 1998.)

Consider an infinite-horizon model, formulated as an optimal control model for social choice and optimal development:

$$\mathbf{V}(a,b) := \text{ MAX } \int_a^\infty e^{-\rho t} f(x(t),u(t))dt \text{ subject to:}$$

$$x(a) = b, \dot{x}(t) = m(x(t),u(t)) \ .$$

Here $x(t)$ is the state function (e.g. capital), $u(t)$ is the control function (e.g. consumption), and $u(t)$ is unconstrained. Denote by $\lambda(t)$ the costate function,

---

9 See also section 2.1.

and by $\Lambda(t) = \lambda(t)e^{\rho\tau}$ the current-value costate function. The current-value Hamiltonian is:

$$\mathbf{h}(x(t), u(t), \Lambda(t)) = f(x(t), u(t)) + \Lambda(t)m(x(t), u(t)).$$

Denote the optimal functions by $x^*(t), u^*(t), \Lambda^*(t)$. Under some restrictions (see Section 3.3), including the requirement (usually fulfilled) that $\lambda(t) = \mathbf{O}(e^{-\beta t})$ as $t \to \infty$ for some $\beta > 0$, and no constraints on the control $u(t)$, it may be shown that:

$$\mathbf{h}(x^*(t), u^*(t), \Lambda * (t)) = -\rho V(t, x^*(t)) = -\rho e^{-\rho\tau} V(0, x^*(t) \to 0 \text{ as } t \to \infty.$$

The proofs in Weitzman (1976), Leonard and Long (1992) and Heal (1998) are incomplete, and do not state some necessary restrictions. Although the two functions:

$$\mathbf{h}(x^*(t), u^*(t), \Lambda * (t)) \text{ and } \delta^{-1} \int_t^\infty f(x^*(\tau), u^*(\tau) e^{-\delta(\tau - t)} d\tau$$

satisfy the same first-order differential equation, their equality only follows if they satisfy a common boundary condition, namely both $\to 0$ as $t \to \infty$, which follows from $\mathbf{h}(x^*(t), u^*(t), \Lambda * (t)) \to 0$.

Weitzman (1976) and subsequent authors have interpreted $V(t, x^*(t))$ as the stock of *total wealth* , and also $\mathbf{h}(x^*(t), u^*(t), \Lambda^*(t))$ as the *interest on total wealth* , which may be taken as a measure of sustainable income, utility or social welfare.

There are many criticisms of this approach to measurement of sustainable welfare and sustainability (Arronsson, Johanssson and Lofgren, 1997,; Brekke, 1997; Heal, 1998; and Land Economics, 1997). The Hamiltonian approach assumes very restrictive and unrealistic conditions, including (i) constant discount rate, technology, and terms of trade, (ii) time autonomous dynamic systems in continuous time, (iii) no constraints on the control or state functions, (iv) positive social time preferences, and (v) convex optimal control models (so that necessary optimality conditions are also sufficient.) The assumption of a single capital good may not be necessary. Some aspects of this approach need more attention, including transversality conditions, and conditions when the Hamiltonian approaches zero for large times. Overtime and intercounting comparisons are not possible; the consumption path has a single peak, and this is not sustained; and the *dictatorship of the present* is embedded in the model. If non-autonomous dynamics are considered, thus assuming $\dot{x}(t) = m(x(t), u(t), t)$ where $m(.)$ depends explicitly on t, then $x(t)$ and $u(t)$ will *not* generally tend to limits $x(\infty)$ and $u(\infty)$ as t tends to infinity; this is the case with the Kendrick-Taylor model considered in Section 5. For an autonomous model, the limits (if they exist) must satisfy $0 = m(x(\infty), u(\infty))$.

Work on computational models and methods for sustainable growth is generally less advanced, although some simple approaches have been tried (see Cesar, 1992; Smulders, 1994; and the journal issue Land Economics, 1997). The object of the present chapter is to develop some mathematical and computational models and methods for sustainable growth, within the framework of welfare economics (Arrow et al., 2003). They include the elements of an optimal growth program (Islam, 2001) such as an objective function, time horizon, time preference and terminal constraints.

There are several ways to embed *sustainability* into mathematical formulations of a computable optimal growth and development models. These include the following:

(*a*) Including environmental factors and consequences in the model, e.g. by dynamic equations, or suitable constraints, for resource supply and exhaustion, pollution generation, etc. (see Heal, 1998; Cesar, 1994). See section 6.5.1.

(*b*) Appropriate specification of the objective function, to express the principles or criteria for sustainable growth (see section 6.3.1)

(*c*) Modelling intergenerational equity, reflected in social time preference, by suitably modified discount factors (see section 6.3.2)

(*d*) Considering a long-term planning horizon, or an infinite horizon in an optimal growth model, so as give suitable weight to the long-term future, as well as to the short-term, e.g. by Rawls (1972) or Chichilnisky (1977). (See sections 6.3.2 and 6.3.3).

The chapter is structured as follows. In sections 6.3 and 6.4, some mathematical models and methods are provided, for including sustainability in an optimal growth model. The Kendrick-Taylor model is modified in Section 6.5 to incorporate sustainability criteria. The results of computational experiments are presented in Section 6.6. Conclusions and computational recommendations are given in section 6.7.

6.3. Modelling sustainability

6.3.1. Description by objective function with parameters

Sustainability should be studied within a social choice framework. The present discussion assumes the possibility of a social choice by a social welfare function (objective function). This function may contain utilitarian and non-welfaristic elements of social choice. It must incorporate, in some suitable form, the concerns for sustainability. If a description using a single objective is sought, that of Chichilnisky (1996) is as plausible as any available. It includes a parameter α, which must be chosen to set the balance between short-term and long-term. This objective function may be written:

$$\alpha \int_0^\infty U(c(t), k(t), t) e^{-\rho t} dt + (1 - \alpha) \lim_{t \to \infty} U(c(t), k(t), t) ,$$

perhaps with some replacement for the discount factor (see Section 4.2). This model assumes that limits exist (so excluding any oscillatory models,) and it is not computable as it stands, because of the infinite time range. As remarked in Section 2, the assumed limits as $t \to \infty$ will normally require that the function m in the dynamic equation, and also the *utility* U, do not depend explicitly on t. Moreover, some bounds on the variables should be adjoined, (compare Rawls, 1972,) and some lower bound on consumption (thus introducing a further parameter.) Note that capital $k(t)$ and consumption rate $c(t)$ are vector functions – though scalar functions may be considered first to simplify the discussion. Note that environmental factors may enter as components of capital; for example, if $p(t)$ is some measure of environmental degradation, then $-p(t)$ could be a component of $k(t)$.

However, a model with a single objective may not represent important aspects of the system being modelled. An alternative approach adjoins constraints, containing parameters. For example, consider a growth model with consumption $u(t)$ and capital stock $k(t)$, and with some floor specified for the consumption. This may relate to a minimum consumption as a function of time, or among various consumers. Some simple examples are as follows.

(a) $\text{MAX}_{u,x}\Phi(x(T))$ subject to $(\forall t)u(t) \geq b_1$,

$$x(0) = x_0, \dot{x}(t) = m(x(t), u(t), t)(t \in [0, T].$$

Here the parameter b describes the minimum consumption level allowed.

(b) $\text{MAX}_{u,x,b}\ b$ subject to $(\forall t)u(t) \geq b_1$,

$$x(0) = x_0, \dot{x}(t) = m(x(t), u(t), t)(t \in [0, T], \Phi(x(T)) \geq \Phi_0.$$

Minimum consumption is maximized, subject to a minimum level for the final capital. A simple special case, with $x(t) \in \mathbf{R}$ and $u(t) \in \mathbf{R}$, is:

$$\text{MAX}_{u,x,b}\ b \text{ subject to } (\forall t)u(t) \geq b_1,$$
$$x(0) = x_0, \dot{x}(t) = \alpha x(t) - u(t)(t \in [0, T], x(T) \geq x_T,$$

at which the optimum (supposing the constraints to be feasible) is evidently at $u(.) = b_1$.

(c) Denoting Pareto maximum by PMAX, consider:

$$\text{PMAX}_{x,u,b}\ \{b, \Phi(x, T))\} \text{ subject to } (\forall t)\ u(t) \geq b,\ b \geq b_0,$$

$$x(0) = x_0, \dot{x}(t) = m(x(t), u(t), t)(t \in [0, T]\ .\ \Phi(x(T)) \geq \Phi_0.$$

(d) A variant of (a) considers a bound on average consumption:

$$T^{-1} \int_0^T c(t)dt \geq b_2$$

(with $b_2 > b_1$ in (a)). This constraint may be included as a penalty cost, by introducing a new state variable $y(t) := \int_0^t T^{-1}c(s)ds$, and then the constraint $y(T) \geq b_2$ enters as a penalty cost of the form $\mu[-y(T) + b_3]_+$ added to the total cost to be minimized. (Here $[.]_+$ replaces negative vector components by zeros, and $b_3 \approx b_2$.)

Note that examples *(a)* and *(b)* will give single elements of the set of Pareto maxima. Several values of the parameter b might be considered, in order to discuss a tradeoff between consumption and capital. A similar remark applies to the parameter α in the Chichilnisky model.

Some other Pareto maxima may be obtained by maximizing a weighted combination $\omega_1^T b + \omega_2 \Phi(T)$. While this is often proposed, some criterion for choosing the weights would be required. It may be more meaningful to consider a parameter such as b instead.

6.3.2. Modified discounting for long-term modelling

A model must maintain a suitable balance between short-term and long-term utility. Consider an optimal control model with a state $x(.)$ (e.g. a capital function) and a control $u(.)$ (e.g. a consumption function), and a time horizon T, with an objective of the form:

$$\int_0^T e^{-\delta t} f(x(t), u(t))dt + \Phi(x(T)).$$

A variant considers $\int_0^\infty$ and $\lim_{t\to\infty} \Phi(x(T)$. The traditional discount factor $e^{-\delta t}$ gives negligible value to the far future, so this is not a sustainable model. If the discount factor is omitted, then the model may not be comparable with models for alternative investment of resources, since it does not allow for possible growth when profits are reinvested.

The $e^{-\delta t}$ discount factor assumes that money can be invested so as to grow at a compound-interest rate $e^{\delta t}$. But such real growth can only happen over a fairly short time horizon, during which suitable investments may be available, perhaps with some limit on the amount invested. The continued exponential growth shown in some datasets for longer time periods merely describes monetary inflation. A continued exponential growth in real terms will soon meet resource constraints (e.g. Forrester, 1971,) or social constraints (e.g. a financial collapse). When a particular activity, or enterprise, is modelled, there is always assumed some *background* of other activities (in the sector, industry, etc.), that is not being described in detail, but rather in some aggregated way. Without a *background* to invest money in, discounting has no meaning

A more realistic discount factor would not assume an indefinite real growth; instead, some saturation effect would appear. A better description is required for *background growth*, meaning the growth available in the larger system, of which the model being studied forms a part. If $\mathbf{g}(t)$ describes the background growth, then a possible description is given by: $\dot{\mathbf{g}}(t) = \delta\mathbf{g}(t)/(1 + \rho t)^2$. This growth starts as exponential, but later approaches saturation. For this growth rate, $\mathbf{g}(t) = \mathbf{g}(0)\ \exp(\delta t/(1+\rho t)$. Thus $\mathbf{g}(t)/\mathbf{g}(0) \to e^{\delta/\rho}$ as $t \to \infty$; qualitatively, saturation starts to matter at about time $1/\rho$. The discount factor is then $\sigma(t) = 1/\mathbf{g}(t)$. Of course, there are other functions than this one with the desired qualitative properties. But, lacking numerical data on saturation, it is appropriate to choose a simple function.

Another possible description for *background* growth assumes a maximum level for capital, and a logistic function for its growth rate, thus:

$$\dot{g}(t) = \delta g(t)[1 - \beta g(t)], g(0) = g_0.$$

This integrates to:

$$g(t) = ae^{\delta t}/[1 + \beta ae^{\delta t}] \text{ where } g_0 = a/(1+\beta a), a = g_0/(1 - \beta g_0).$$

Thus, as $t \to \infty$, g(t) $\to 1/\beta$, the assumed maximum level. This model may be more plausible than the previous one where the saturation is a given function of time t. The discount factor is then $\sigma(t) := 1/g(t) = e^{-\delta t} + \beta a$. Thus it tends to a positive constant value for large times.

Heal (1998, page 98) considers a class of discount factors $\Delta(t)$ satisfying $\Delta'(t)/\Delta(t) \to 0$ as $t \to \infty$, for which the term $\lim_{t\to\infty}\Phi(x(T)$ does not affect the optimum. The classical exponential discount factor does not satisfy this condition; the modified discount factor $1/\mathbf{g}(t)$given above satisfies it. Then Heal's criterion is satisfied for both the functions σ (.) proposed here, since:

$$\dot{\sigma}(t)/\sigma(t) = -\delta/(1+\rho t) + \delta\rho/(1+\rho t)^2 or - \delta/(1 + \beta ae^{\delta t},$$

and each of these $\to 0$ as $t \to \infty$.

The Strotz phenomenon (see Chakravarty 1969, page 41) shows that only with an exponential discount factor can a certain balance be achieved between an optimum path starting at time 0 and an optimum path starting at a later time. Unfortunately this is incompatible with giving proper weight to the more distant future.

6.3.3. Infinite horizon model[10]

Optimal control models of the form:

$$\text{MIN}_{u(.),x(.)}F(x,u) := \int_0^\infty e^{-\rho t} f(x(t), u(t), t)dt \text{ subject to:}$$

$$x(0) = x_0, \dot{x}(t) = m(x(t), u(t), t),\ \text{a(t)}\ \leq u(t) \leq b(t)(t \geq 0),$$

have often been considered in recent economic literature (sometimes as MAX $F(x,u) \Leftrightarrow$ MIN $-F(x,u)$), often without explicit bounds on the control $u(t)$, and often with $e^{-\rho t}$ as the only explicit dependence on t. However, for the

[10] See also Section 2 2 and chapter 9.

infinite horizon, it may not be obvious whether a minimum, or maximum, is reached; and the Pontryagin theory of optimal control is commonly presented only for a finite time horizon, since some assumptions about uniform approximation over the whole time domain are involved. For an infinite horizon, some further restrictions are required for validity. The present discussion assumes limiting behaviour as $t \to \infty$, which ensures that the infinite-horizon problem is closely approximated by a problem with finite horizon T. Consequently, any weight attached to the distant future must be expressed by a separate term in the objective, as e.g. in the Chichilnisky model.

Assume that a minimum is reached, at $(x^*(t), u^*(t))$. The discount factor $e^{-\rho t}$ is needed, so that $F(x, u)$ is finite. Following the approach in Craven (1995), the differential equation with initial condition is expressed in abstract form as $Dx = M(x, u)$; the Hamiltonian is:

$$h(x(t), u(t), t, \lambda(t) := e^{-\rho t} f(x(t), u(t), t) + \lambda(t) m(x(t), u(t), t),$$

and the integral of the Hamiltonian equals $H(x, u, \theta) := F(x, u) + \theta M(x, u)$, where θ is a Lagrange multiplier vector, and $\lambda(t)$ is a function representing θ.

Assume that $x^*(t), u^*(t)$, and $\lambda * (t)$ (from the adjoint equation below) tend to limits as $t \to \infty$, with $\lambda * (t) = \mathbf{O}(e^{-\beta t})$ as $t \to \infty$; $f(., ., .)$ and $m(., ., .)$ are twice differentiable functions, such that the second derivatives:

$$f_{xx}(x(t), u(t), t), m_{xx}(x(t), u(t), t), f_{xu}(x(t), u(t), t), m_{xu}(x(t), u(t), t)$$

are bounded in t, whenever $(x(t), u(t))$ are near to $(x^*(t), u^*(t))$.

Note that an oscillatory optimum control is thus excluded; and the exponential decay terms mean that the infinite horizon is closely approximated by problem over (0, T), for some suitable finite T.

The steps in the proof, based on (Craven, 1995) are as follows; only the significant changes for an infinite horizon need be detailed.

(a) Obtain first-order necessary conditions.

(b) From (a), deduce the adjoint differential equation:

$$-\dot{\lambda}(t) = (\partial/\partial x) h(x(t), u(t), t, \lambda(t)) \ , \ \lambda(t) \to 0 \text{ as } t \to \infty.$$

Here $\lambda(t)$ is assumed to represent θ, subject to verification that the differential equation obtained is solvable; the calculation using integration by parts is valid, under the assumptions on limits as $t \to \infty$.

(c) A linear approximation:

$$H(x, u, \theta) - H(x^*, u, \theta) = H_x(x^*, u^*, \theta*)(x - x^*) + \mathbf{o}(\|u - u^*\|)$$

holds, given the bounded second derivatives, noting that a term $e^{-\rho t}$ is present in F_x, and $e^{-\beta t}$ in θM_x; here $\|u - u^*\| = \int_0^\infty |u(t) - u^*(t)| dt$. The *quasimin* property follows:

$$H(x, u^*, \theta) - H(x^*, u^*, \theta) \geq \mathbf{o}(\|u - u^*\|).$$

(d) Assume that control constraints, if present, are a constraint on $u(t)$ for each time separately. If $h(x^*(t), \; . \; , t, \lambda * (t))$ is *not* minimized at $u^*(t)$ (possibly except for a set E of t for which $\int_E dt = 0$), then a standard proof (not depending on the domain of t) shows that the quasimin is contradicted.

Suppose now that an endpoint term $\lim_{T \to \infty} v(x(T))$ is added to the objective function. This has the effect of adding $v(x(t))\delta(t - T)$ to f, then letting $T \to \infty$. In consequence, the boundary condition for the adjoint equation becomes:

$$\lambda(t) - v'(x(t)) = \mathbf{O}(e^{-\beta t}) \text{ as } t \to \infty.$$

If $x(t)$ tends to a limit as $t \to \infty$ (which implicitly assumes that f and m do not explicitly contain t), and $v(.)$ is continuous, then the limit of $v(x(T))$ exists. This remark applies to the Chichilnisky criterion in Section 6.4.2.

6.4. Approaches that might be computed

6.4.1. Computing for a large time horizon

As discussed in Section 3.1, it is appropriate to discuss optimization models with parameters – required to take account of the several conflicting objectives that exist – and the question of what happens to the optima when these parameters are varied provides a further stage of exploration. However, the model must be somewhat recast, in order that a computer package can handle it, whether RIOTS_95 (Schwarz, 1996) or any other. Assuming that the limits as $t \to \infty$ exist, a *nonlinear time* transformation $t = \psi(\tau)$ may be applied, so as to compress the time scale for large times t (when little is changing). A computation requires a discretization of the time interval into finitely many subintervals, and such a transformation considerably reduces the number of subintervals required (see Craven, 1995).

6.4.2. The Chichilnisky criterion with a long time horizon

If a discount factor $e^{-\rho t}$ is retained, then the Chichilnisky model may be approximated, for some large T, by:

$$\text{MAX}_\alpha \int_0^T e^{-\rho t} U(c(t), k(t)) dt + \alpha U(c(T), k(T)) \rho^{-1} e^{-\rho T}$$

$$+ (1 - \alpha) V(k(T))$$

$$\text{subject to } k(0) = k_0; \;\; \dot{k}(t) = m(c(t),$$

$$k(t)), c_U \geq c(t) \leq c_L (0 \leq t \leq T); \;\; k(T) \geq k_T.$$

The second term is an estimate of the contribution for times beyond T; a similar estimate is the *salvage value* cited in (Chakravarty, 1969). The following assumptions are made:

- the utility $U(.,.)$ and the dynamics function $m(.,.)$ have no explicit dependence on time t;
- T is assumed large enough that $c(T)$ and $k(T)$ approximate their limits as $\rightarrow \infty$;
- the term in $e^{-\rho t}$ estimates the integral from T to ∞;
- the lower bound c_L on consumption is assumed constant (achievable by scaling $c(t)$;
- the upper bound c_U is included to bound the region where an algorithm must search;
- the utility V applicable at time T may, but need not, be the same as U.

Note that the utility V depends on the final capital $k(T)$, but not on the final consumption $u(T)$, since $u(T)$ is not of significance, but rather the whole curve of consumption versus time.

The constraint $k(\infty) \geq k_T$ may be replaced by a penalty term:

$$\frac{1}{2}\mu[k(\infty) - k_T + \mu^{-1}\epsilon]_+^2 \ ,$$

where μ is a positive parameter, and $[\ ]_+$ replaces negative components by zero; $\epsilon \geq 0$ will be adjusted, so as to fulfil the constraint exactly. According to the theory of *augmented Lagrangians* (see the discussion in Craven, 1978), ϵ relates to a Lagrange multiplier. If $c(t)$ does not meet its bounds, then the inactive constraints on $c(.)$ can be omitted. According to the Pontryagin theory, with some regularity assumptions because of the infinite domain (notably that the limits exist), an optimum satisfies the differential equation for $\dot{k}(t)$, the adjoint differential equation:

$$-\dot{\lambda}(t) = -e^{-\rho t}U_k(c(t), k(t)) + \lambda(t)m_k\ (\mathrm{c(t),\ k(t),\ t}),$$

$$\lambda(\infty) = \mu[k(\infty) - k_T + \mu^{-1}\epsilon]_+ - (1 - A)V_k(k(\infty);$$

and (from the Pontryagin principle):

$$-\alpha e^{-\rho t}U_c(c(t), k(t)) + \lambda(t)m_k(c(t), k(t), t)(0 \leq t < T).$$

The adjoint differential equation is required for computing the gradient of the objective.

When T is large, and the only explicit time dependence is the discount factor, it may be useful to transform the time scale nonlinearly by:

$$t = \psi(\tau) := -\rho^{-1}log(1 - \beta\tau), \text{ where } \beta = 1 - e^{-\rho T} \ ;$$

this maps $\tau \in [0,1]$ to $t \in [0, T]$; $dt/d\tau = (\beta/\rho)(1-\beta\tau)^{-1} = (\beta/\rho)e^{\rho t}$. Denoting $u(\tau) := c(\psi(\tau))$ and $x(t) := k(\psi(\tau))$, the problem is transformed to:

$$\mathrm{MAX}J(u) := \alpha \int_0^1 U(u(\tau), x(\tau))d\tau + \alpha U(u(1), x(1))\rho^{-1}e^{-\rho T}$$

$$+(1-\alpha)V(u(1), x(1))$$

$$= \alpha \int_0^{1[} U(u(\tau), x(\tau)) + \delta(\tau - 1)W(u(\tau), x(\tau)]d\tau$$

$$\text{where } W(u(\tau), x(\tau) := U(u(\tau), x(\tau)\rho^{-1}e^{-\rho T} + \alpha^{-1}(1-\alpha)V(x(\tau))$$

$$\text{subject to: } x(0) = k_0;$$

$$\dot{x}(\tau) = (\beta/\rho)m(u(\tau), x(\tau), \psi(\tau))/(1-\beta\tau)(0 \le t \le T),$$

$$c_U \ge u(t) \ge c_L(0 \le \tau \le 1);\ \mathrm{x}(1)\ \ge k_T.$$

Here $\dot{x}(\tau)$ means $(d/d\tau)x(\tau)$.

The adjoint differential equation is then:

$$-\dot{\lambda}(t) = -\alpha U_x(u(\tau).x(\tau)) + (\beta/\rho)m_x(u(\tau), x(\tau), \psi(\tau))/(1-\beta\tau);$$

$$\lambda(1) = -(\alpha/\rho)e^{-\rho T}U_x(u(1), x(1)) - (1-\alpha)V_x(u(1), x(1)) + \mu[k(T) - k_T$$

$$+\mu^{-1}\epsilon]_+$$

The gradient of the objective is then computed (when $z(1) = 0$) from :

$$J'(u)z = -\alpha \int_0^T [U_u(x(\tau), u(\tau)) + (\beta/\rho)\lambda(\tau)m_u(u(\tau), x(\tau))/(1-\beta\tau)]z(\tau)d\tau$$

6.4.3. Chichilnisky model compared with penalty term model

In the approximated Chichilnisky model considered in Section 6.4.2, the second term with $e^{-\rho T}$ estimates $\int_T^\infty e^{-\rho t}$ U(c(t), k(t))dt, in the case when $c(T)$ and $k(T)$ are close to limiting values. (For the Kendrick-Taylor model discussed below in Section 5, there are no limiting values.) The comparison for a discount rate $\rho = 0.03$ and horizon $T = 20$ years):

$$\int_0^T e^{-\rho t}dt = \rho^{-1}(1 - e^{-\rho T}) \approx\ 0.55 \text{ with } e^{-\rho T} \approx\ 15.04$$

shows that the term is unimportant, if U and V are of comparable size.

Neglecting it, the objective reduces to:

$$\alpha\{\int_0^T e^{-\rho t}U(c(t), k(t))dt + \alpha^{-1}(1-\alpha)V(k(T))\}.$$

If $k(.)$ has only one component, then a reasonable choice for $V(k(T))$ would be:

$$-\xi[k(T)-k^{\#}]^2 \text{ or } -\xi[k(T)-k*]_+^2 ,$$

where μ is a suitable positive parameter. (Of course, another form than quadratic could be chosen; the quadratic is convenient to compute). The two versions describe attainment of a target, or perhaps exceeding it. Then the objective becomes:

$$\alpha\{\int_0^T e^{-\rho t}U(c(t),k(t))dt + \frac{1}{2}\mu[k(T)-k^{\#}]^2\},$$

with $\mu = 2\xi\alpha^{-1})1-\alpha)$, and perhaps [ ] replaced by $[\,]_+$.

This may be compared with adding to the integral a penalty term, to represent an endpoint constraint $k(T) = k_T$ (or $k(T) \geq k_T$). The penalty term has the same form, with k* differing a little from k_T, the difference depending on μ and on a Lagrange multiplier in the optimization. In fact, the SCOM package handles an endpoint constraint in exactly this way.

It follows that the modified Chichilnisky model may be studied as a parametric problem, with μ and k* as parameters.

6.4.4. Pareto optimum and intergenerational equity

Intergenerational equity (and also sustainability) can be interpreted in terms of Pareto optimality. Consider now two objective functions, say:

$$F^1(x,u) := \int_0^T e^{-\rho t} f(x(t),u(t))dt \text{ and } F^2)x,u) := \Phi(x(T))$$

to describe, in some sense, utilities for the present generation and a future generation, where $x(t)$ denotes rate of consumption, and $k(t)$ denotes capital (both may be vectors). Assume that these functions are constrained by a dynamic equation for $(d/dt)x(t)$, and bounds on $u(t)$. Suppose that (x^*,u^*) is a *Pareto maximum* point of this model. Then (assuming some regularity of the constraint system), Karush-Kuhn-Tucker necessary conditions, or equivalent Pontryagin conditions, hold for (x^*,u^*) exactly when these conditions hold for a single objective function $\tau_1 F^1(.) + \tau_2 F^2(.)$, for some nonnegative multipliers τ_1, τ_2, not both zero. Different points in the (large) set of Pareto optima correspond to different choices of the multipliers. For the two-objective problem, the costate becomes a matrix function $\Lambda(t)$, and the Pontryagin maximum principle considers a Pareto maximum of a vector Hamiltonian:

$$(F^1(x,u), F^2(x,u)) + \Lambda(t)(RHS \text{ of dynamic equation })$$

with respect to $u(t)$ (see Craven, 1999).

This may be compared with the parametric version of Chichilnisky's criterion:

$$\alpha F^1(x,u) + (1-\alpha)F^2(x,u) = \alpha[F^1(x,u) + \beta F^2(x,u)], \beta = \alpha^{-1}(1-\alpha).$$

Each choice of α, or β, gives a different Pareto maximum point.

6.4.5. Computing with a modified discount factor

If the discount factor is modified to $e^{-\rho t} + \kappa$, then integration over an infinite time domain will give infinite values. Instead, the following model may be considered:

$$\text{MAX } \alpha \int_0^T [e^{-\rho t} + \kappa] U(c(t), k(t)) dt + (1 - \alpha) V(k(T))$$

subject to $k(0) = k_0; \dot{k}(t) = m(c(t), k(t), t), c_U \geq c(t) \geq c_L (0 \leq t \leq T); k(T) \geq k_T$.

Here, $\kappa > 0$, and some finite time horizon T must replace ∞. With the same time transformation as above (for computation), the transformed problem becomes:

MAX $\alpha \int_0^1 (1 + \kappa(1 - \beta\tau)^{-1}) U(u(\tau), x(\tau)) d\tau + (1 - \alpha) V(u(1), x(1))$

$$= \alpha \int_0^1 (1 + \kappa(1 - \beta\tau)^{-1}) U(u(\tau), x(\tau)) + \delta(\tau - 1)\alpha^{-1}(1 - \alpha) V(x(\tau))] d\tau$$

subject to $x(0) = k_0; \dot{x}(\tau) = (\beta/\rho) m(u(\tau), x(\tau), \psi(\tau))/(1 - \beta\tau)(0 \leq \tau \leq 1)$.

6.5. Computation of the Kendrick-Taylor model

6.5.1. The Kendrick-Taylor model

The Kendrick-Taylor model for economic growth (Kendrick and Taylor, 1971), which is a Ramsey type (Ramsey, 1928) model, has the form:

$$\text{MAX}_{c(.), k(.)} \int_0^T a e^{-\rho t} c(t)^\theta \text{ dt subject to}$$

$$k(0) = k_0, \dot{k}(t) = \zeta e^{\delta t} k(t)^\beta - \sigma k(t) - \text{c(t)}, \ k(T) = k_T.$$

Here $c(t)$ denotes consumption, and $k(t)$ denotes capital (including man-made, natural, and environmental, and human). Some computational results for this model are given in Craven and Islam (2001) and Islam and Craven (2001).

Since the model includes a growth factor $e^{\delta t}$, it may be appropriate (see Chakravarty, 1969) to choose the terminal value k_T to increase exponentially with T . If $k(t) = k_0 e^{\omega t}$ and $c(t) = c_0 e^{\omega t}$, then the dynamic equation is satisfied only when $\omega = \delta/(1-\beta)$ and $k_0\omega = k_0^\beta - \sigma k_0 - c_0$. Then the objective function remains finite as $T \to \infty$ when $\rho > \theta\omega$, thus when $\rho > \theta\delta/(1-\beta)$. (An analogous criterion is given by Chakravarty (1969, page 99) for a different model, only partly described.) Since the growth of $c(t)$ is dominated by the discount factor, the optimal objective tends to a limit as $T \to \infty$. So here an infinite-horizon model may be approximated by a finite-horizon model; but it does not describe sustainable growth over an infinite horizon, since the exogenous growth factor $e^{\delta t}$can only be sustained for a limited time.

6.5.2. Extending the Kendrick-Taylor model to include a long time horizon. [11]

This model is now modified to use the modified discount factor from section 6.3.2. This gives the formulation:

$$\mathrm{MAX}_{c(.),k(.)} \int_0^T a[e^{-\rho t} + \kappa]c(t)^\theta \text{ dt subject to}$$

$$k(0) = k_0, \dot{k}(t) = \zeta e^{\delta t} k(t)^\beta - \sigma k(t) - \text{ c(t)}, \; k(T) = k_T.$$

This formulation differs from the model in Kendrick-Taylor (1971) only by the inclusion of the positive parameter κ . In order to compare with previous results, the following numerical values are considered:

$$a = 10, T = 10, \rho = 0.03, \theta = 0.1, f10k_0 = 15.0, \zeta = 0.842,$$

$$\beta = 0.6, \sigma = 0.05, k_T = 24.7.$$

However, larger values of the horizon T become relevant. The parameter κ depends on the relative weighting to be given to the longer-term in relation to the short-term. Since $e^{-\rho t}$ is small when $t \geq 2/\rho$, the ratio of the two could be taken as:

$$\int_{2/\rho}^T \kappa dt \; / \int_0^{2/\rho} e^{-\rho\tau} dt \approx \kappa\rho T.$$

A possible value for this parameter would be 0.5. Otherwise:

$$\int_0^T \kappa dt = \int_0^T e^{-\rho t} dt$$

when $\kappa \approx 0.86$.

In a more general model, $c(t)$ and $k(t)$ would take vector values; however, single components for $c(t)$ and $k(t)$ will be considered here.

With the modified discount factor $e^{-\rho t} + \kappa$, integration over an infinite time domain will give infinite values. So the horizon T must here be finite (though it may be large). An alternative to the terminal constraint $k(T) = k_T$ is a terminal objective term, giving an objective function:

$$\int_0^T a[e^{-\rho t} + \kappa]c(t)^\theta dt + \varphi(c(T)).$$

[11] See also section 2.2 and chapter 9.

With the same time transformation as in section 6.4.2, and writing $\tilde{c}(\tau) = c(t)$ and $\tilde{k}(\tau) = k(t)$, the objective function and dynamic equation become:

$$(\beta/\tau)\int_0^1 a[1+\kappa/(1-\beta\tau)]\tilde{c}(\tau)^\theta d\tau + \varphi(\tilde{c}(1))$$

$$\tilde{k}(0) = k_0, \dot{\tilde{k}}(\tau) = (\beta/\rho)[\zeta e^{\delta t}\tilde{k}(\tau)^\beta - \sigma\tilde{k}(t) - \tilde{c}(t)]/(1-\beta\tau).$$

6.5.3. Chichilnisky variant of Kendrick-Taylor

The Chichilnisky formulation does not directly apply to the Kendrick-Taylor model, because $\lim_{t\to\infty} x(t)$ is not available. However, the modified version of Section 4.2, with $e^{-\rho t}$ discount and some large horizon T, may be considered for the Kendrick-Taylor model. If the endpoint constraint is replaced by a penalty cost $V(x(T)) := [x(T) - k_T]^2$, then the nonlinear time transformation leads to:

$$U(u(\tau), x(\rho)) = u(\tau)^\theta; W(u(\tau), x(\tau)) = u(\tau)^\theta \rho^{-1} e^{-\rho T}$$

$$+\alpha^{-1}(1-\alpha)[x(t) - k_T]^2;$$

$$x(0) = k_0, \dot{x}(\tau) = (\beta/\rho)[\zeta(1-\beta\tau)^{-\delta/\rho}x(\tau)^\beta - \sigma x(\tau) - u(\tau)]/(1-\beta\tau).$$

However, because of the growth term $e^{\delta t}$ in the dynamic equation, there is no infinite-horizon version of the Kendrick-Taylor model, and the time transformation was not obviously useful here. Computations omitting this growth term may be of interest - see Section 6.6.4.

6.5.4. Transformation of the Kendrick-Taylor model

In view of some numerical instability encountered when computing the Kendrick-Taylor model for time horizons $T > 10$, the following transformation (see Islam and Craven, 1995) of the model to an equivalent, more computable, form may be used. Set $q(t) = e^{\sigma t}k(t)$ and $\theta = \sigma + \delta - \beta\sigma$; then:

$$\dot{q}(t) = e^{\sigma t}(\zeta e^{\delta t}k(t)^\beta - \sigma k(t) - u(t) + \sigma k(t));$$

$$\dot{x}(t) = \gamma\zeta e^{\theta t}(x(t))^{(\beta+\gamma-1)/\gamma} - \gamma u(t),$$

$$\text{where}\quad u(t) = e^{\sigma t}q(t)^{\gamma-1}c(t) = e^{\sigma t}(x(t))^{(\gamma-1)/\gamma}c(t).$$

Choosing $\gamma = 1 - \beta$, the problem in the new functions $x(t)$ and $u(t)$ becomes:

$$\text{MIN}\int_0^T ae^{-\nu t}u(t)^\epsilon x(t)^\mu dt \text{ subject to:}$$

$$\dot{x}(t) = \gamma\zeta e^{\theta t} - \gamma u(t)\ , \ x(0) = k_0^\gamma, x(T) = k_T^\gamma,$$

where $\nu = \rho + \epsilon\sigma$ and $\mu = \epsilon(1-\gamma)/\gamma.The$ dynamic equation now does not involve fractional powers. Note that this transformation does *not* preserve any bounds on $c(t)$.

Using the data from section 6.5.2, with $k()) = 15.0$,

$$\gamma = 0.4, \nu = 0.035, \mu = 0.15, x(0) = 2.9542, \theta = 0.04$$

If $T = 10, k(0) = 15.0, k(T) = 24.7$, then $x(T) = 4.4049$. If $T = 20$ then $k(T)$ may be considered as $15.0 + 2(24.7 - 15 - 0) = 34.4$, then $x(T) = 6.1424$.

6.6. Computer packages and results of computation of models

6.6.1. Packages used

Optimal solutions for these sustainable growth models may be computed, using appropriate computer packages for optimal control. The packages used were SCOM (Craven and Islam, 2001), and also RIOTS_95 (Schwartz et al., 1997), to validate the SCOM results by comparison with another package. In the SCOM package, the control function is approximated by a step-function, constant on each of N subintervals of the time period, with e.g. $N = 20$; this is known (Craven, 1995) to be an adequate approximation. Function values and gradients are computed, by solving differential equations; then the MATLAB *constr* package for mathematical programming is used to compute the optimum. For the models considered here, the gradients given by the theory were not useful (see discussion in Section 8), and finite-difference approximations were provided by *constr*. Another suitable package for optimal control is OCIM (see Craven, de Haas and Wettenhall, 1998).

For analysing the results of all model computations, we have followed the common practice in economic growth economics, where the arbitrary effects of terminal constraints are avoided by ignoring the results of the last time periods, thus from when the variables start to tend towards the terminal constraints (see for example Land Economics Journal, 1997)

6.6.2. Results: comparison of the basic model solution with results for modified discount factor.

The following computations were done with various modifications of the Kendrick-Taylor model.

Results computed with the SCOM package were obtained for the discount factor $e^{-\rho t} + \kappa$ for capital (upper graph of Figure 1) and consumption (lower graph), for a ten-year time horizon, for the three cases:

(a) $\kappa = 0.00$ Objective = 98.1 ;

(b) $\kappa = 0.86$ Objective = 196.0 ; and

(c) $\kappa = 1.72$ Objective = 243.9.

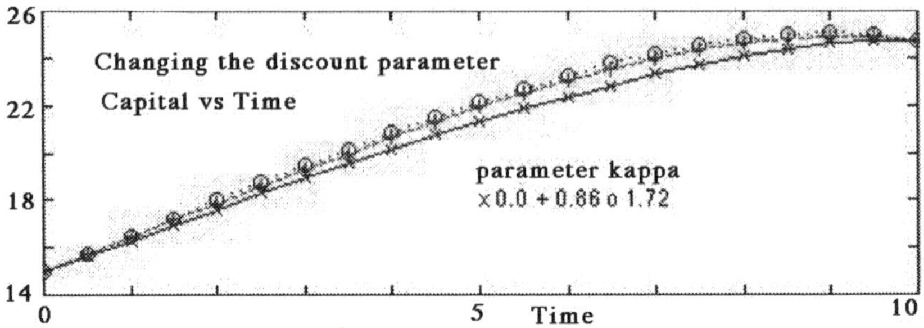

Fig. 1a

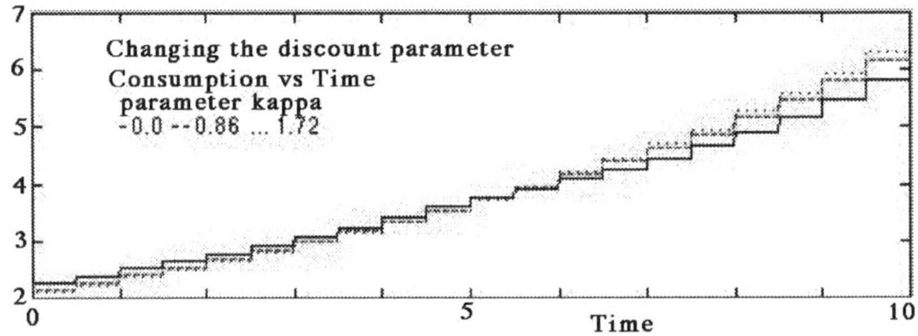

Fig. 1b

Within each graph, (a) is the lowest curve, and (c) is the highest. Note that the objective values are not comparable, since the different discount factors measure utility on different scales.

In order to assess the accuracy of these computations, Fig. 2 compares the computed consumptions computed by the SCOM and the RIOTS_95 packages. The following values were obtained for for objective functions:

$\kappa =$	.0	0.1	0.8639	1.7278
RIOTS_95 calculation	98.08	109.46	196.41	294.74
SCOM calculation	98.08		195.96	293.85

The results for the two packages are in substantial agreement. As remarked in Craven and Islam (2001a), the optimal curve of capital is insensitive to small rapid fluctuations in the consumption (the control function), so a step-function approximation to the control is sufficient. Increasing the parameter kappa in the modified discount factor, so as to give more weight to later times, decreases the consumption for earlier times and increases it for later times, and also increases the rate of capital growth.

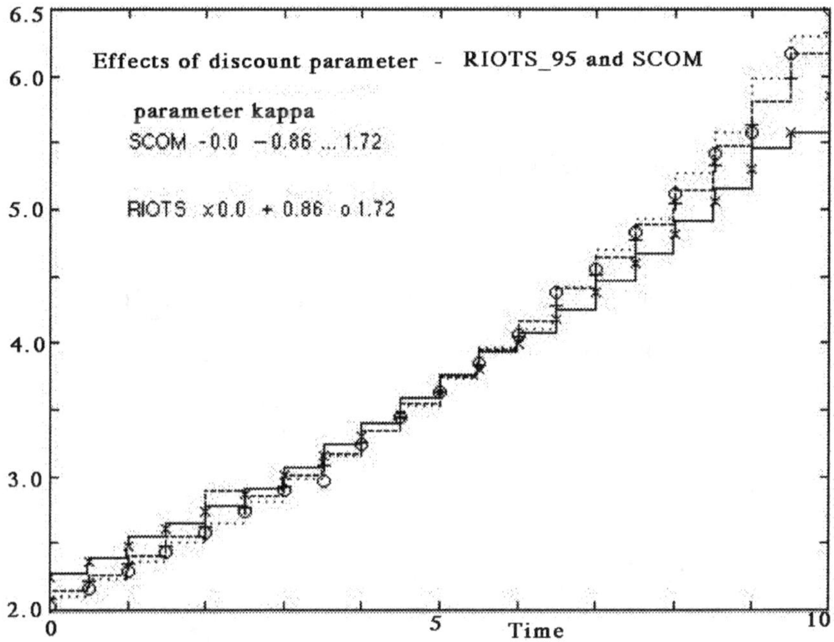

Fig. 2

6.6.3. Results: Effect of increasing the horizon T[12]

Figures 3 and 4, computed by SCOM using the transformation of the Kendrick-Taylor model from section 5.4, show capital and consumption for a horizon of T increases from 10 to 20 years, and with the final capital value $k(T)$ increased from $k(10) = 24.7$ to $k(20) = 34.4$ (obtained from $k(0)+2(24.7-$ k(0)). For comparison, results from RIOTS_95 are also plotted.

Consider the numerical values:

$$a = 10, \rho = 0.03, \epsilon = 0.1, \zeta = 0.842, \delta = 0.02,$$

$$\sigma = 0.05, k_0 = 15,0, T = 20, k_T = 34.4.$$

Then:

$$\gamma = 0.4, \nu = 0.035, \mu = 0.15, x(0) = 2.9542, x(T) = 4.1174 \ , \ \theta = 0.04 \ .$$

[12] See also section 5.7.

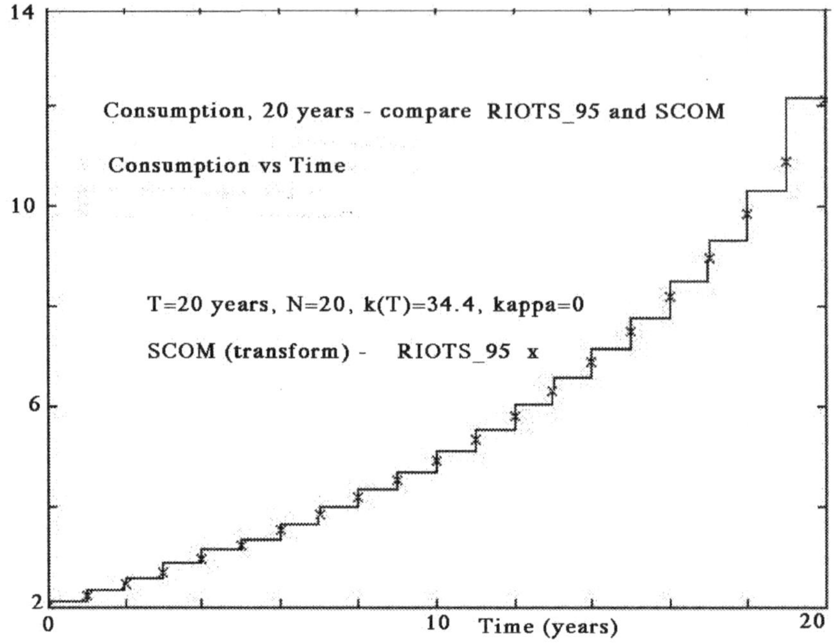
Consumption, 20 years - compare RIOTS_95 and SCOM
Consumption vs Time
T=20 years, N=20, k(T)=34.4, kappa=0
SCOM (transform) - RIOTS_95 x
14
10
6
2
0
10
Time (years)
20

Fig. 3

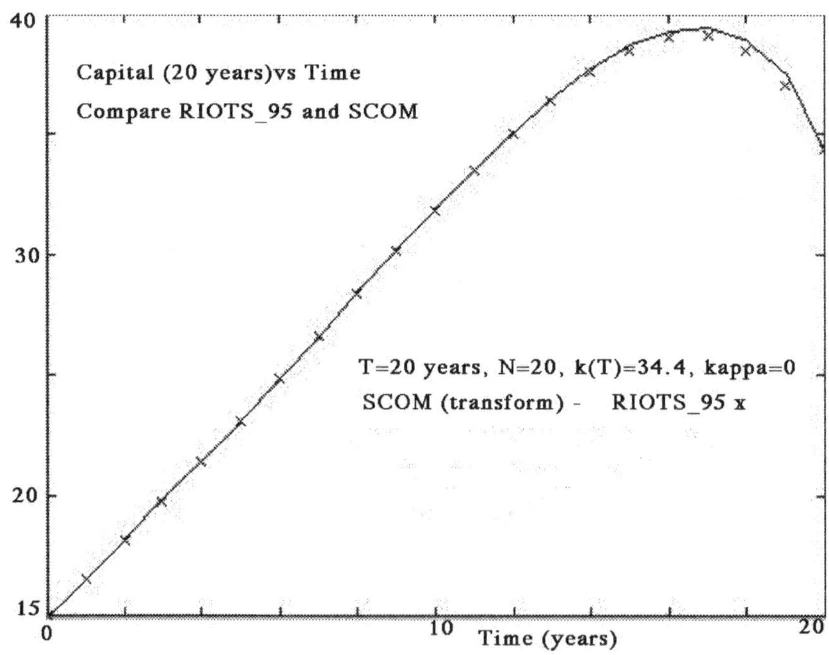
Capital (20 years)vs Time
Compare RIOTS_95 and SCOM
T=20 years, N=20, k(T)=34.4, kappa=0
SCOM (transform) - RIOTS_95 x
40
30
20
15
0
10
Time (years)
20

Fig. 4

Figures 5 and 6 (computed with RIOTS_95) show the result of increasing the parameter κ in the modified discount factor. As with a ten year horizon, the consumption is decreased for earlier times and increased for later times. The capital growth is increased, then brought down to the target of $k(20) = 34.4$.

6.6.4. Results: Effect of omitting the growth term in the dynamic equation.

The result of making $\delta = 0$ in the growth term $e^{\delta t}$ is shown in Figures 7 and 8.

As a check on accuracy, Figures 9 and 10 show the result of increasing the number N of subdivisions from 20 to 40 (for the case $\kappa = 0, T = 20, k(T) = 34.4$, growth term present). The calculations agree closely.

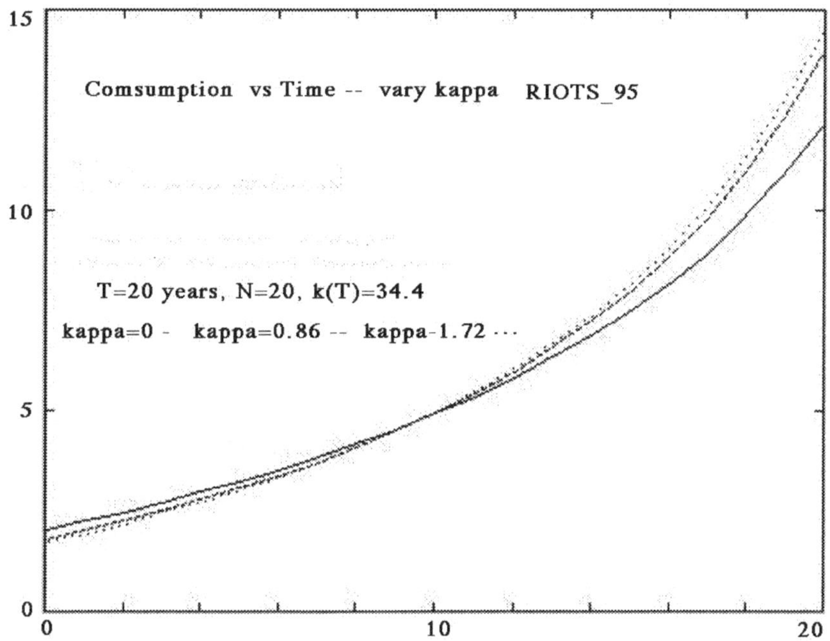

Fig. 5

6.6.5. Results: Parametric approach[13]

Figures 11 and 12 show the effect of changing the parameter $k(T)$, the specified capital at the final time $T = 20$ of the calculation. The discount parameter κ was kept as 0. The final capital $k(T)$ could be brought up to 55, with some loss in consumption.

The effect of changing the exponent β in the dynamic equation was studied in Islam and Craven (2001a).

[13] For parametric approach, see also section 6.4.2.

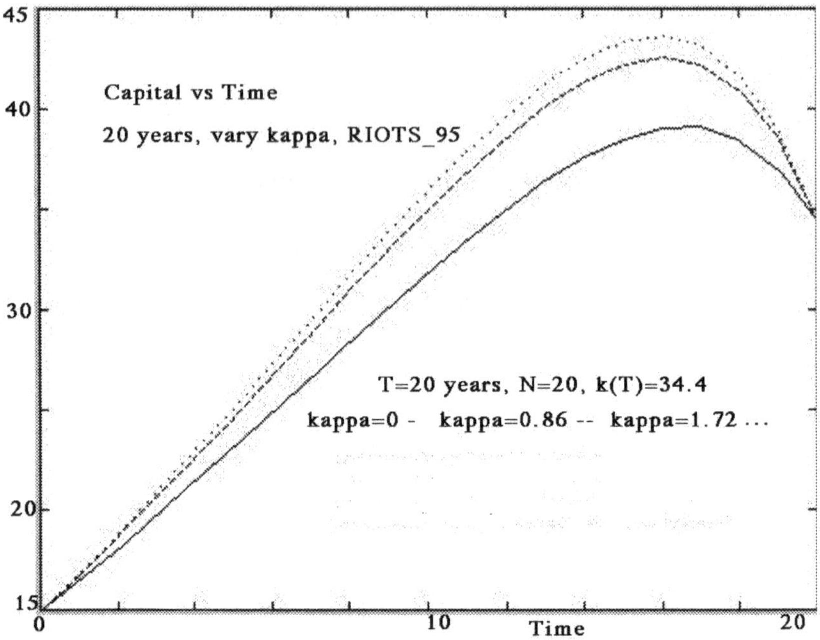
Capital vs Time
20 years, vary kappa, RIOTS_95
T=20 years, N=20, k(T)=34.4
kappa=0 - kappa=0.86 -- kappa=1.72 ...
45
40
30
20
15
0
10
Time
20

Fig. 6

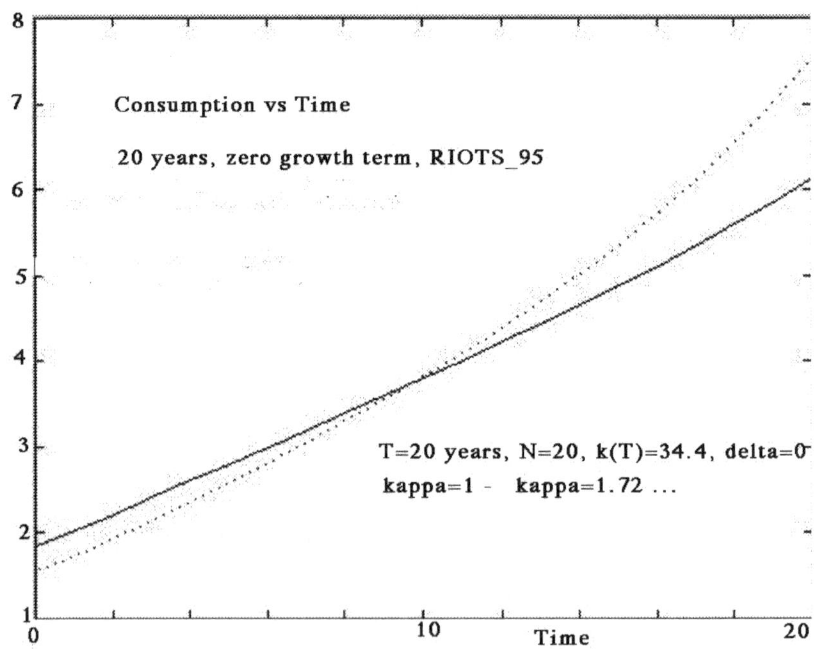
Consumption vs Time
20 years, zero growth term, RIOTS_95
T=20 years, N=20, k(T)=34.4, delta=0
kappa=1 - kappa=1.72 ...
8
7
6
5
4
3
2
1
0
10
Time
20

Fig. 7

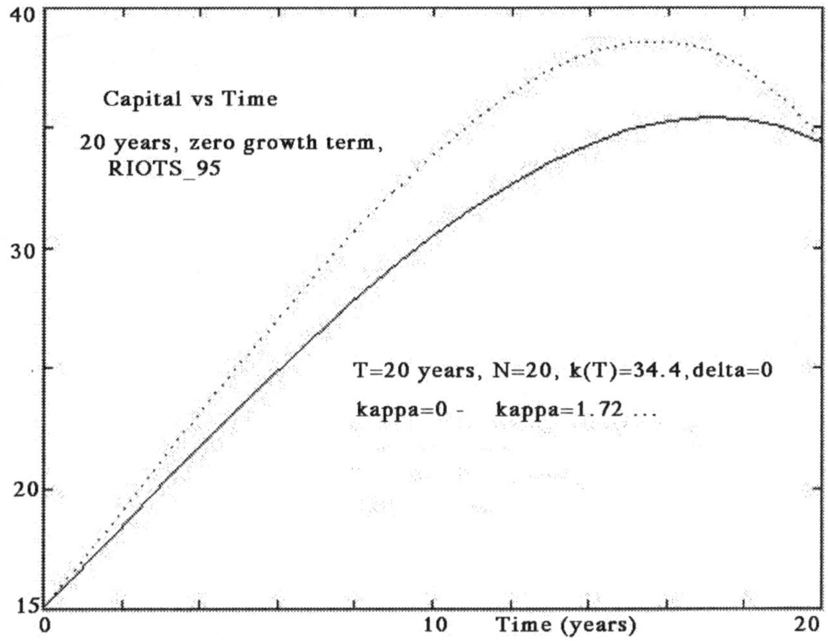
Capital vs Time
20 years, zero growth term, RIOTS_95
T=20 years, N=20, k(T)=34.4,delta=0
kappa=0 - kappa=1.72 ...
40
30
20
15
0
10
Time (years)
20

Fig. 8

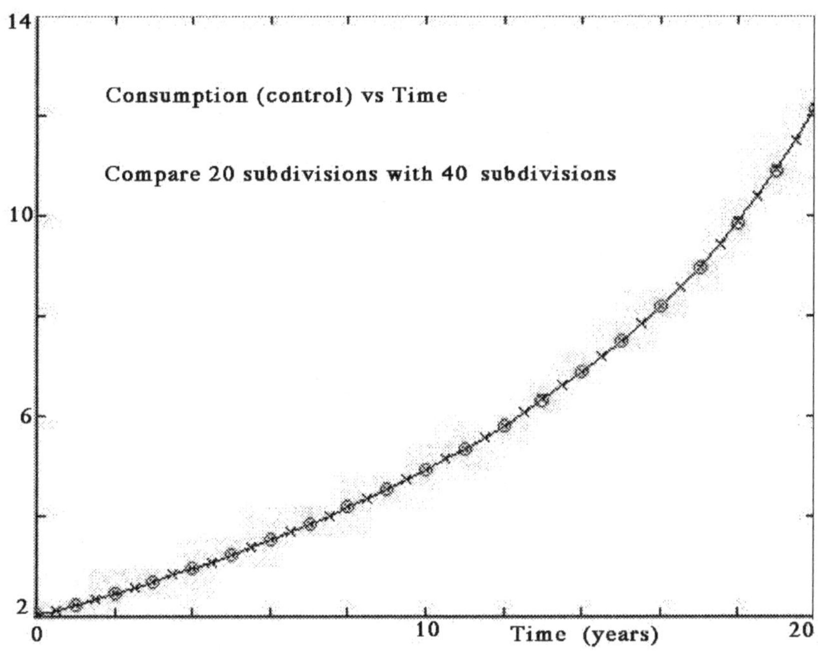
Consumption (control) vs Time
Compare 20 subdivisions with 40 subdivisions
14
10
6
2
0
10
Time (years)
20

Fig. 9

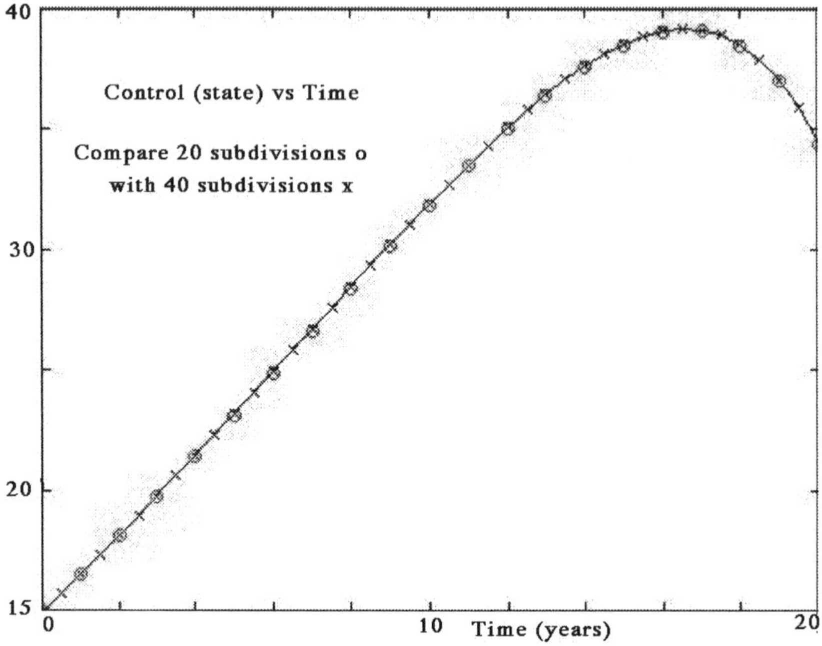
Control (state) vs Time
Compare 20 subdivisions o
with 40 subdivisions x
40
30
20
15
0
10
Time (years)
20

Fig. 10

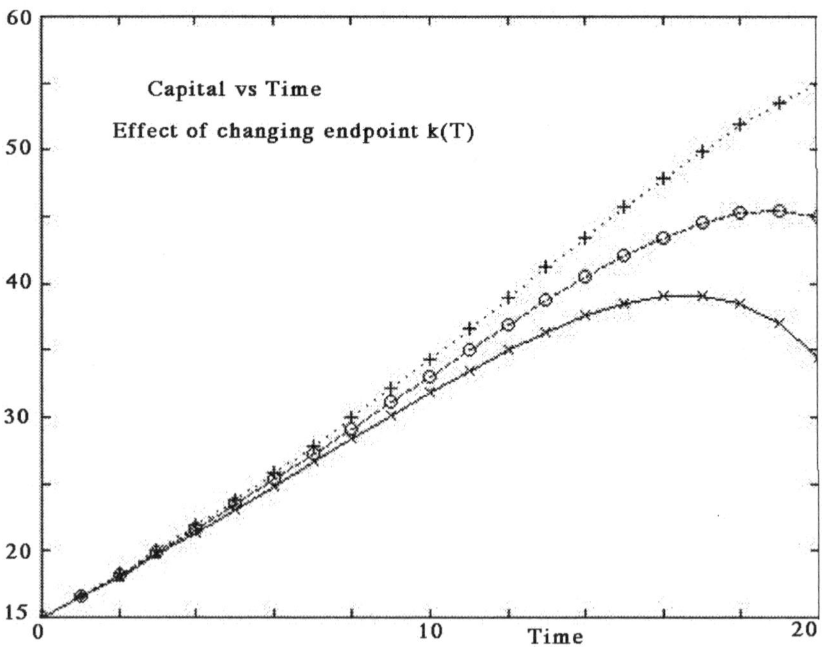
Capital vs Time
Effect of changing endpoint k(T)
60
50
40
30
20
15
0
10
Time
20

Fig. 11

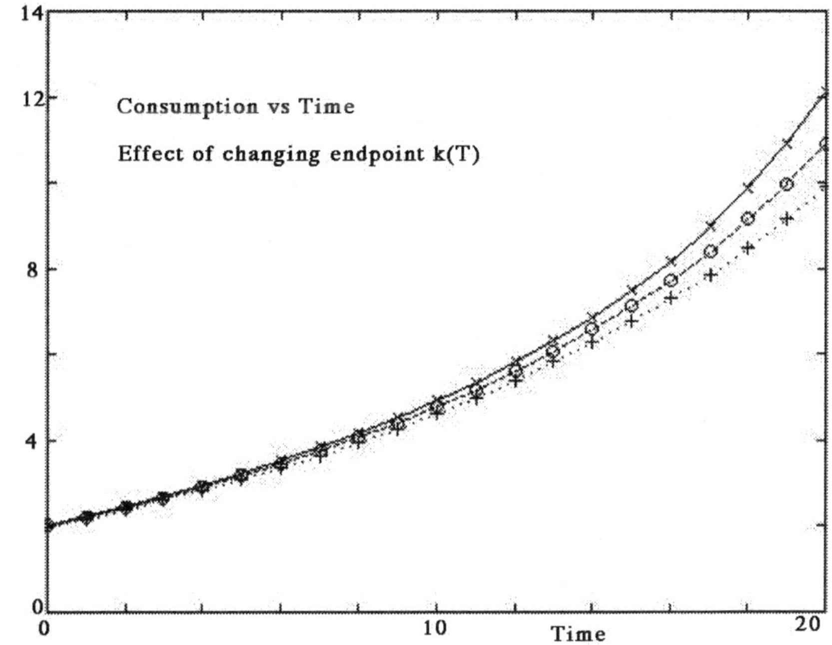

Consumption vs Time
Effect of changing endpoint k(T)
14
12
8
4
0
0
10
Time
20

Fig. 12

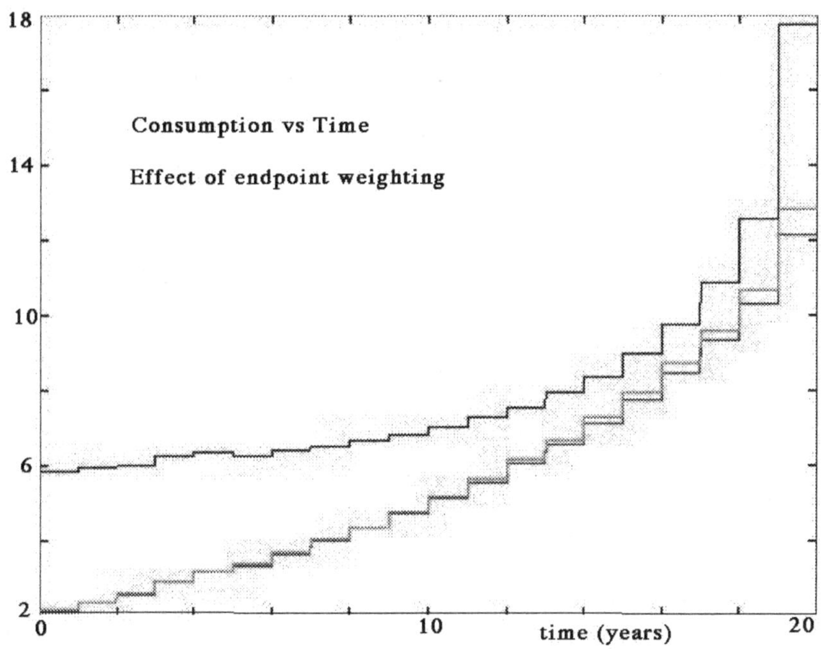

Consumption vs Time
Effect of endpoint weighting
18
14
10
6
2
0
10
time (years)
20

Fig. 13

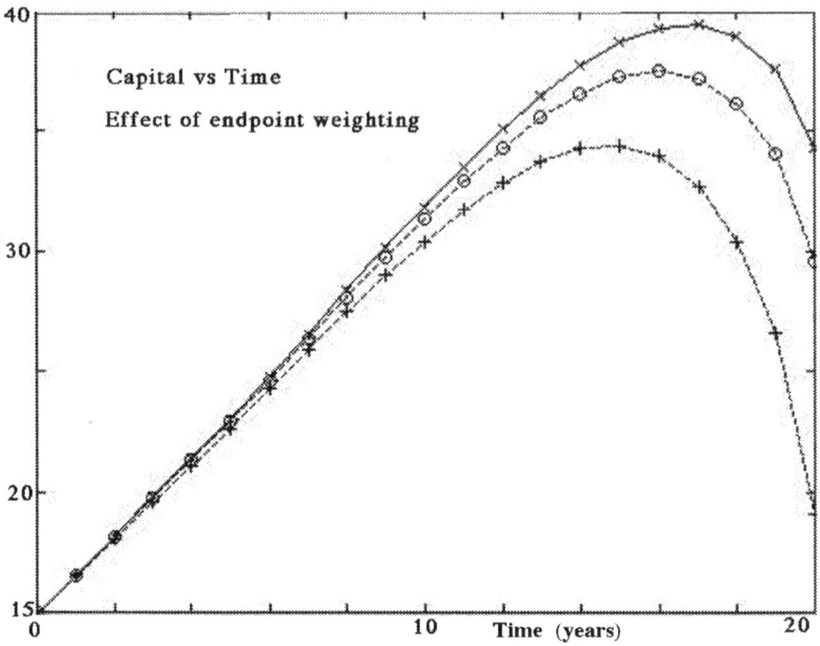

Fig. 14

6.6.6. Results: the modified Chichilnisky approach

The modified Chichilnisky approach of Sections 4.2 and 4.3, with a large but finite horizon T, may be applied to the Kendrick-Taylor model. As discussed in 4.3, the second term of the objective may be neglected. So this modified Chichilnisky problem may be studied as a parametric problem, with μ and k* as parameters. Figures 13 and 14 show consumption and capital, computed by SCOM for the Kendrick-Taylor model with horizon $T = 20$ years (and the discount parameter $\kappa = 0$), with the endpoint term included, and the target $k(T) = 34.4$. The three cases are:

$\mu = 10$ (lower graph:); $\mu = 0.1$ (middle graph); $\mu = 0.02$ (upper graph).

The case $\mu = 10$ was used previously in SCOM to reach the target $k(20) = 34.4$ accurately.

Clearly, decreased weight for the endpoint term reduces capital, and increases consumption.

6.7. Existence, uniqueness and global optimality

In an optimal control model for an economic growth problem, there is usually no way to prove existence in advance. The following procedure deals with this question.

Assuming (for the moment) that an optimum exists, one may first construct Pontryagin necessary conditions, then find a solution to these necessary conditions. This solution is not always an optimum; however it is an optimum if the problem satisfies convex, or *invex*, conditions (See Craven, 1995). The Kendrick-Taylor model is certainly not convex, but does satisfy *invex* conditions (see Islam & Craven, 2001). Therefore the solution obtained to the Pontryagin conditions is indeed an optimum.

6.8. Conclusions

Several social choice models of sustainability in optimal growth models are described, with computational methods. Two optimal control packages were used, Schwarz's RIOTS_95 and Craven's SCOM, to test and validate the results. The model results have a turnpike property, as shown in (Islam and Craven, 2001). The computed results show time paths of economic variables which show relatively higher values of consumption or welfare during the later part of the planning horizon. The results for the two computing packages were in good agreement.

Computations with versions of the Kendrick-Taylor growth model presented some difficulties, because of implicit constraints in the model (see Craven and Islam, 2001). An optimization calculation must compare various consumption functions, and some produce negative values for capital, so an optimum is not reached. This difficulty arose for a 20-year horizon, and was avoided with the RIOTS_95 package by computing using finite differences for gradients, and with SCOM by a mathematical transformation of the model to an equivalent and more stable form.

From a comparison of the original form of the Kendrick-Taylor model with other, more sustainable, forms, there are some implications as to sustainable growth. Effects of changing the discount factor to give more weight to the distant future, of omitting the exogenous growth factor in the dynamic equation, and of varying the target for endpoint capital, are presented in the graphs. While the Chichilnisky model is not directly computable because it involves an infinite time horizon, a modified version with a large finite horizon can be computed. This version is equivalent to attaching a penalty cost when a capital target is not reached. The solutions so obtained are Pareto maxima of a two-objective problem, considering both a short-term objective and a long-term objective.

The computational experiments in this chapter, using several different models for optimal growth, show that sustainability of growth and social welfare can conveniently be modelled empirically for policy analysis, and which specify growth paths that are sustainable and ensure intergenerational equity, altruism and fairness.

6.9 User function programs for the transformed Kendrick-Taylor model for sustainable growth

```
    % (Gradients not supplied)
function ff=yj(t,it,z,yin,hs,um,xm)
                        % Integrand 20e-.75tu(t)0.1x(t)0.15
ff=-20.0*exp(-0.705*t)*(um(floor(it),1)ˆ0.1)*(iqq(xm,hs,t)ˆ0.15);
function ff=yx(t,it,z,yin,hs,um,xm)
                        % (d/dt)x(t) = 20(0.842e0.8t - 0.4u(t)
ff=20*(0.842*0.4*exp(0.8*t) - 0.4*um(floor(it),1));
function ff=yf(xf,um,xm)  % Penalty term 10(x(1) - 6.124)∧2
ff=10*(xf-6.1424)∧2;
function gg=yc(ii,hs,um,xm)
gg=0;
```

Chapter 7

Modelling and Computing a Stochastic Growth Model

7.1. Introduction

An economy progresses in an environment that is characterised by uncertain static and dynamic forces and processes. Uncertain factors which underlie the dynamic growth processes correspond to market prices, ecosystem, technical progress, population growth, fluctuations in output, exports, imports, exchange rates, and economic and social factors. Social choice in an uncertain environment is more problematic and controversial than in a certain environment (Islam, 2001). Models which incorporate these uncertain factors in economic growth and development are useful for understanding uncertain processes of economic dynamics. A suitably specified stochastic model can be used to discuss social choice in uncertain conditions. This requires specification of a decision criterion, and an appropriate mathematical description of the uncertainty. This is feasible and plausible with the assumptions of new[3] welfare economics. An earlier work on modeling stochastic growth is by Haavelmo (1954). Other contributions include Aoki (1989), Kendrick (1981), Tintner and Sengupta (1969), and Sengupta and Fanchon (1997). Taylor and Uhlig (1990) provide an up to date exposition of methods for modeling non-linear stochastic optimal growth economics, applied to a particular computational model. Other references related to stochastic optimization in economics include Arrow (1951, 1971), Islam (1999, 2001a), Kelly and Kolstad (1997), Radner (1982), Tapiero (1998).

Economic models including stochastic elements have been studied either (a) in continuous time, or (b) in discrete time. For continuous time, the stochastic elements are usually described by a Wiener process, for which the stochastic contributions from nonoverlapping time intervals are independent, however short the intervals. This may not happen in the real world. A discrete-time model, say with time intervals $\delta > 0$, avoids this assumption, but may require extensive calculations of probability distributions, e.g. for a Markov chain. However, if the stochastic elements are not too large, it may be enough to compute the mean and variance, instead of the whole distributions.

A continuous-time deterministic model for economic growth and development may be approximated by a discrete-time model, by discretizing the time. If small stochastic terms are then added to this discrete-time model, dynamic (difference) equations may be obtained for the mean and variance. These lead, in turn, to continuous-time dynamic equations for mean and variance. They relate to the original model, with small stochastic terms added, but now only assuming independence for contributions over non-overlapping time intervals of length δ.

To explore this approach, a small stochastic term is added to the classical (deterministic) Kendrick-Taylor model for economic growth (Kendrick and Taylor, 1971; Kendrick, 1981), considered as an optimal control model for computation (see Islam and Craven, 2000). Other classical growth models (see Chakravarty, 1969; Fox, Sengupta and Thorbecke, 1973) could be similarly modified by adding a stochastic element. The computations were done using the RIOTS_95 package (see Schwartz, Polak and Chen, 1997), using MATLAB (see MATLAB, 1997). Other optimal control packages that might have been used include SCOM (Islam and Craven, 2000), and MISER3 (Jennings, Fisher, Teo and Goh, 1991).

7.2. Modelling Stochastic Growth

Consider a dynamic model in continuous time, described by:

$$x(0) = x_0, \ \dot{x}(t) = m(x(t), u(t), t) + \theta(t) \quad (0 \leq t \leq T),$$

with a stochastic term $\theta(t)$. Assume only that the stochastic contributions for successive time intervals $(j\delta, (j+1)\delta)$, $j = 1, 2, \ldots$ are independent. Here $x(\cdot)$ is the state function, whose components may include capital, and terms describing resource utilization and environmental factors, and $u(\cdot)$ is the control function, which may describe consumption. Note that $\theta(\cdot)$ is not assumed to be a Wiener process. Also, the stochastic terms can, in practice, be only known approximately; in particular, the distribution is never known precisely, and usually only an approximate variance. So it may suffice to calculate the stochastic term rather approximately, while the expectation (= mean) is calculated more precisely.

Now discretize the time, with time intervals δ, to obtain an approximating discrete-time process for $z(j) := x(\delta j$ and $v(j) := v(\delta j$ as:

$$z(j+1) - z(j) = m(z(j), v(j), \delta j) + \xi(j),$$

where the stochastic terms $\xi(j)$ are now assumed independent. If the $\xi(j)$ are not too large, the mean and variance of $z(j)$ can be sufficiently described by difference equations. For convenient computation, the difference equations may be approximated by differential equations in continuous time t. The result is a good approximation to the mean of $x(t)$, and a useful approximation to the variance of $x(t)$, for the model described.

Other authors have considered approximating such models by discretizing in various ways the state space or the time. Taylor and Uhlig (1990) have surveyed a number of such approaches for a discrete-time model of the form:

$$x_t = \rho x_{t-1} + \epsilon_t,$$

where the random variables ϵ_t are uncorrelated normal with mean zero and constant variance. The approaches include a *value-function grid,* where time and the state variables are discretized, *quadrature value-function grid,* which differs by using a quadrature rule to discretize the state space, approximation by *linear-quadratic* control problems, a method involving first-order conditions and conditional expectations, and an *Euler-equation grid.* Haavelmo (1964) (see also Fox, Sengupta and Thorbecke (1973)) has also considered the model (3), and has given conditions for the variance of the state variable to tend to infinity, or to tend to a limiting value. These are special cases of the discrete-time stochastic model of Amir (1997), where the distribution of x_t is described by a transition probability, satisfying certain convex or nondecreasing requirements. The optimum is described by a functional equation of dynamic programming type. If the contributions ϵ_t are assumed independent, then the process is Markovian, and could be described by a probability distribution for each time t. For the case where the stochastic contributions are relatively small compared to the deterministic terms, these distributions could be approximated, at each time t, by computing only their mean and variance.

7.3. Calculating mean and variance

Consider the Kendrick-Taylor dynamic model:

$$x(0) = x_0, \quad \dot{x}(t) = 0.842e^{.02t}x(t)^{0.6} - u(t) - 0.05x(t) \equiv \psi(x(t), u(t))$$

over time horizon $t \in [0, T]$, with $T = 10$ and $x_0 = 15.0$ Subdivide $[0, T]$ into N equal subintervals, say with $N = 20$. A discrete-time approximation is given by:

$$x((j+1)T/N) \approx x(jT/N) + \psi(x(jT/N), u(jT/N)) \equiv \varphi(x(jT/N), u(jT/N))$$

in which $u(jT/N)$ means $u((jT/N) + 0)$ if u has a jump at jT/N. Consider then the model:

$$x((j+1)T/N) = \varphi(x(jT/N), u(jT/N)) + y(j),$$

where $y(j)$ is a stochastic variable with mean 0 and variance v, with $y(j)$ independent of $x(j)$, and $y(1), \ y(2), \ \ldots$ distributed independently.

For given $u(.)$, approximate $\phi(.)$ by $k + ax(.) + bx(.)^2$ over a suitable range of $x(.)$, say $[15, 25]$ for the numbers used in (1). Note that a and b do not depend on $u(.)$. Suitable values are $a = 0.2541$ and $b = -0.0018$ .Then the expectation $x_1(.)$ and variance $x_2(.)$ of $x(.)$ are described by:

$$x_1((j+1)T/N) = \varphi(x_1(jT/N) + bx_2(jT/N);$$

$$x_2((j+1)T/N) \approx (a^2 + 4abx_1(jT/N)x_2(jT/N) + v.$$

The equation for expectation is approximate, but may be sufficiently accurate when the stochastic terms are small. It is based on the following calculation. If X and Y are stochastic variables, and $Z = k + aX + bX^2$, then Z has expectation and variance given by :

$$\mathbf{E}Z = k + a\mathbf{E}X + b(\mathbf{E}X)^2 + b\mathbf{var}X;$$

$$\mathbf{var}Z = \mathbf{var}Y + (a^2 + 5b^2(\mathbf{E}X)^2 + 4ab\mathbf{E}X))\mathbf{var}X$$
$$+(4b^2\mathbf{E}X + 2ab)\mathbf{E}(X^3) + b^2\mathbf{var}(X^2).$$

Then the equation for $\dot{x}_2(t)$ is obtained by omitting some small terms, assuming b is not too large. Note that, as j increases, the variance $x_2(.)$ is increased by the successive addition of terms $v = \mathbf{var}Y$, and also decreased because the power 0.6, reflected in the negative coefficient b, reduces the contribution of $\mathbf{var}x(t)$

These difference equations are now replaced by differential equations (in continuous time):

$$\dot{x}(t) = \psi(x_1(t), u(t)) + (N/T)bx_2(t);$$

$$\dot{x}_2(t) = (Nv/T) + (N/T)(a^2 - 1 + 4abx_1(t))x_2(t).$$

The rough approximation used here only affects the stochastic terms, whereas the calculation of expectation is more precise. The objective function to be maximized is:

$$J(u) = \int_0^{10} 10.0(e^{-0.03t} + \kappa)u(t)^{0.1}\, dt,$$

where κ is zero for the original Kendrick-Taylor model, and positive if greater weight is given to the more distant future. The state function $x(.)$ does not enter this objective function, so it is not necessary to consider its expectation.

This optimal stochastic growth model requires specification of a number of parameters, including the rate of time discounting, the parameter κ describing the weight to be given to the more distant future, the time horizon (taken here as $T = 10$ years), the endpoint condition for the capital $x(T)$, and the variance parameter v.

7.4. Computed results for stochastic growth

The computed results demonstrate the intertemporal optimal allocation of resources in uncertain conditions (*stochastic social choice*). The following diagrams show the expected capital (Figure 1), the variance of the capital (Figure 2), and the consumption (Figure 3), as functions of time, with different values of the parameter v, describing the variance of the stochastic term added to the dynamic equation, and also when the endpoint constraint on the capital at the end time T is changed. Here $\kappa = 0.0$, as in the original Kendrick-Taylor model. Figures 4,5,6 show comparable results when $\kappa = 0.8639$, a value considered to describe greater weight given to times after the horizon $T = 10$.

When the variance parameter v is changed, the expected capital will change slightly, but this change is too small to show on the graphs. Comparison of Figure 1 with Figure 4 shows, as expected, that the capital rises higher when a greater weight is given to the future. The variance of the capital (Figure 2) increases considerably as v increases. For later times, the variance is strongly influenced by the endpoint condition imposed on the expected capital $x(T)$ at time T. Thus, a turnpike effect is not apparent on this time scale; it may appear with a larger time horizon. The effect of endpoint is greater when the future has more weight (Figure 5). When the endpoint condition is $x(T) = 45$, the computed variance in Fig. 5 runs slightly negative; the approximation in the model is not quite adequate here.

The graphs of consumption (Figure 3 and Figure 6) show very little effect of the variance parameter v, within the range considered. As expected, the consumption decreases somewhat when the endpoint condition $x(T)$ increases. When the future has more weight (Figure 6), the consumption is generally higher at later times.

Figures 7,8 and 9 show what happens to capital, variance of capital, and consumption when the variance parameter v is increased over a greater range. Such an increase (from $v = 0$, thus the deterministic model, up to $v = 40$) increases the expected capital (Figure 7) a little (except near the horizon $T = 10$ years where the expected capital is fixed by an endpoint condition), increases the variance of capital considerably (Fig. 8), and decreases the consumption a little (Figure 9). The changes are continuous as the variance parameter v increases up from zero. The growth of capital may be visualized by a band:

$$\text{Expected Capital } \pm 2\sqrt{(\text{Variance of Capital})},$$

containing about 95% of the probability distribution, instead of just a curve.

The objective function for this model happens to depend on the control function (consumption), but not on the state function (capital); so there is no need to consider here the expectation of the capital.

The results are consistent with the consensus in the literature (see Smulders, 1994) on the effects of uncertainty on economic growth.

Over the range of parameters considered here, the consumption and the expectation of capital are not greatly influenced by stochastic terms. The expected capital, its variance, and consumption change continuously as the small stochastic term increases from zero; the graphs include the deterministic case (no added stochastic term), for comparison. The expectation of capital grows along a unique stable path. For the tine horizon (10 years) considered, the endpoint condition on the expectation of capital has a strong effect.

In Islam and Craven (2003c), the existence and global uniqueness of the optimum for the Kendrick-Taylor model were established, using a mathematical transformation that converts the problem to an equivalent convex formulation (see also Islam, 1999). This transformation does not extend to the stochastic model studied here. However, the (deterministic) Kendrick-Taylor model has a strictly stable optimum, and the stochastic model considered is a small perturbation of the deterministic model. Hence, by Craven (1995), Theorem 4.7.1, an optimum exists for the stochastic model. However, this is not quite enough to prove uniqueness, though this is likely.

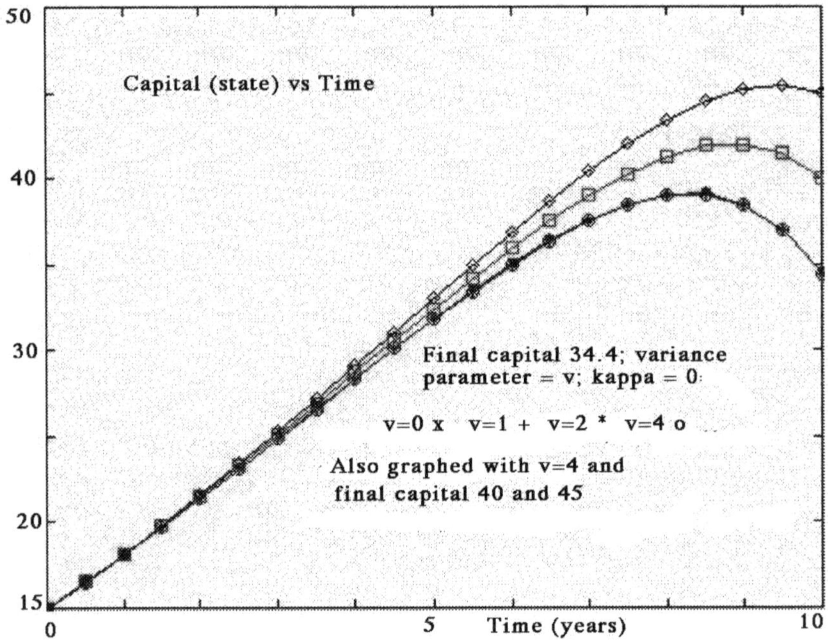

Fig. 1

7.5. Requirements for RIOTS_95 M-files

Compared with the usual (deterministic) Kendrick-Taylor model, the number of states is increased from 1 to 2. The M-files *sys_h* for the right hand side of the dynamic equation, and *sys_dh* for its gradient, take the following forms:

$$\text{xdot} = [0.842 * \exp(0.02 * t) * x(1) \wedge 0.6 - u(1) - 0.05 * x(1) \ldots$$

$$-2.0 * 0.0028 * x(2); \ldots$$

$$\mathbf{2.0} - 2.0 * (0.9354 + 0.00183 * x(1)) * x(2)];$$

(Here ... is MATLAB's code for "continued next line".)

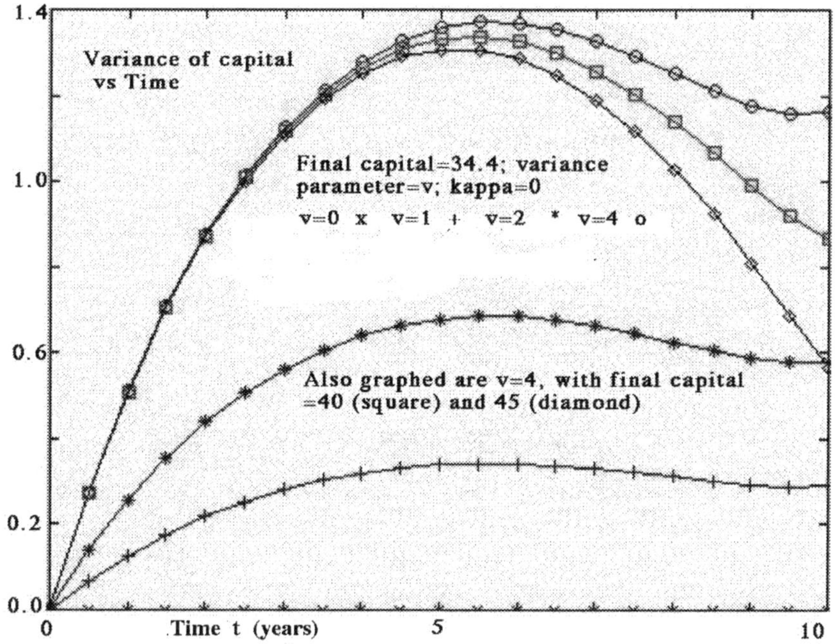
Variance of capital
vs Time
Final capital=34.4; variance
parameter=v; kappa=0
v=0 x v=1 + v=2 * v=4 o
Also graphed are v=4, with final capital
=40 (square) and 45 (diamond)
1.4
1.0
0.6
0.2
0.0
0
Time t (years)
5
10

Fig. 2

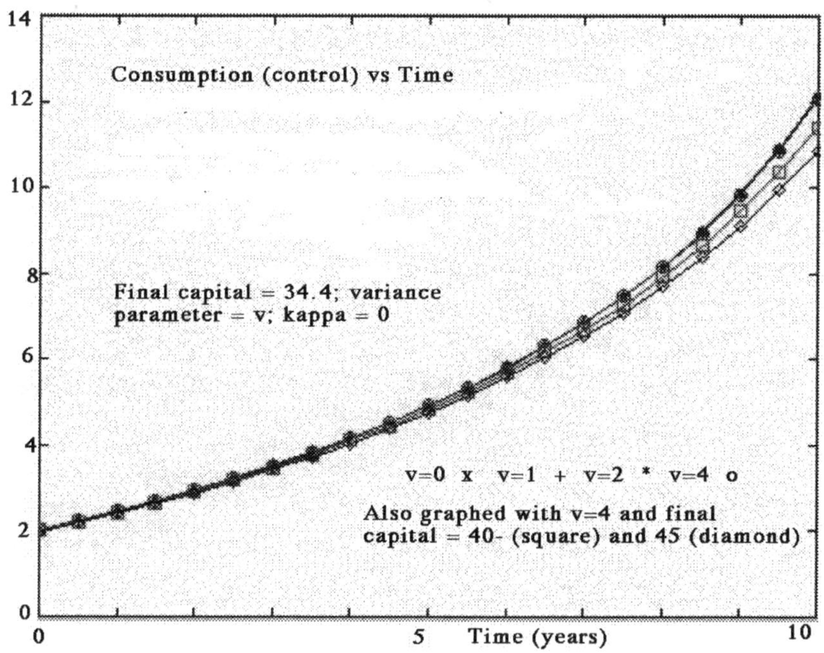
Consumption (control) vs Time
Final capital = 34.4; variance
parameter = v; kappa = 0
v=0 x v=1 + v=2 * v=4 o
Also graphed with v=4 and final
capital = 40- (square) and 45 (diamond)
14
12
10
8
6
4
2
0
0
5
Time (years)
10

Fig. 3

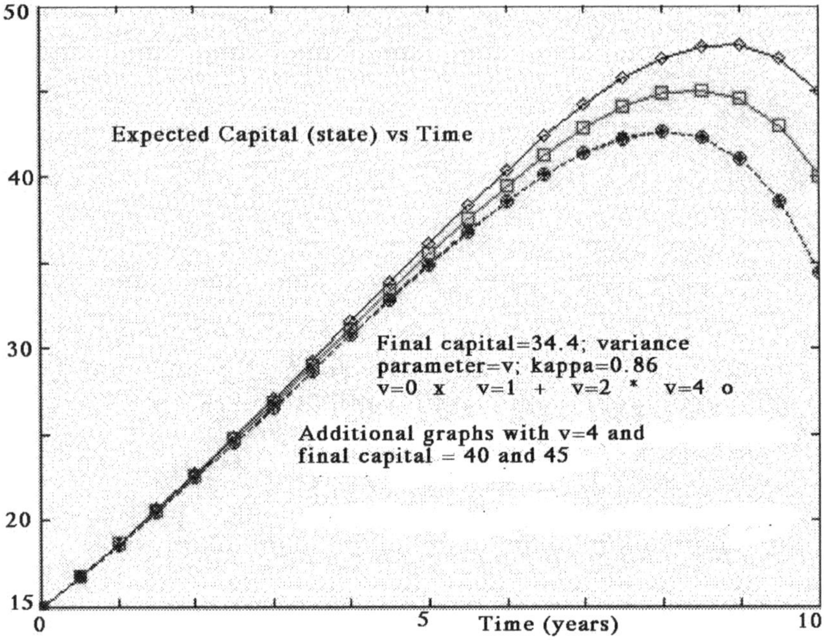

Expected Capital (state) vs Time
Final capital=34.4; variance parameter=v; kappa=0.86
v=0 x v=1 + v=2 * v=4 o
Additional graphs with v=4 and final capital = 40 and 45
Time (years)

Fig. 4

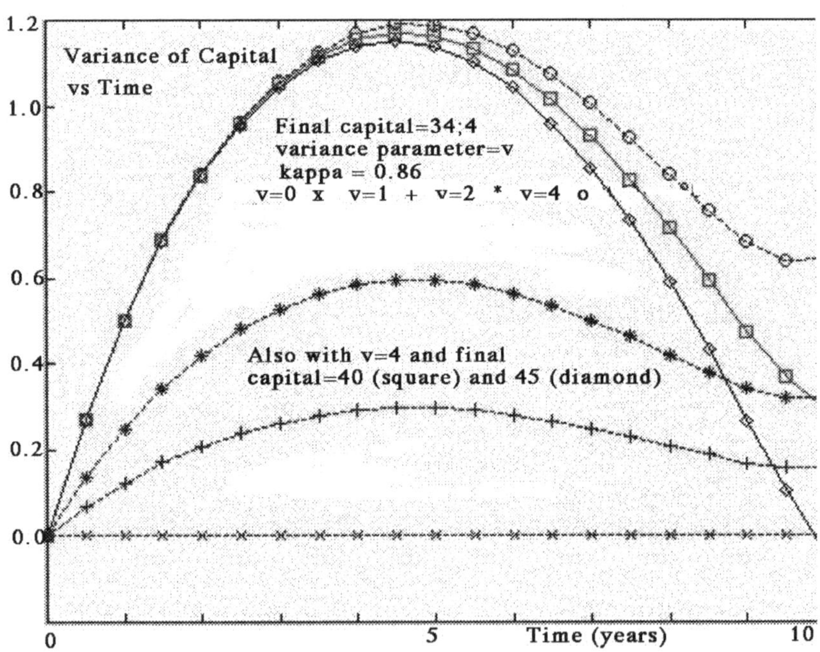

Variance of Capital vs Time
Final capital=34;4
variance parameter=v
kappa = 0.86
v=0 x v=1 + v=2 * v=4 o
Also with v=4 and final capital=40 (square) and 45 (diamond)
Time (years)

Fig. 5

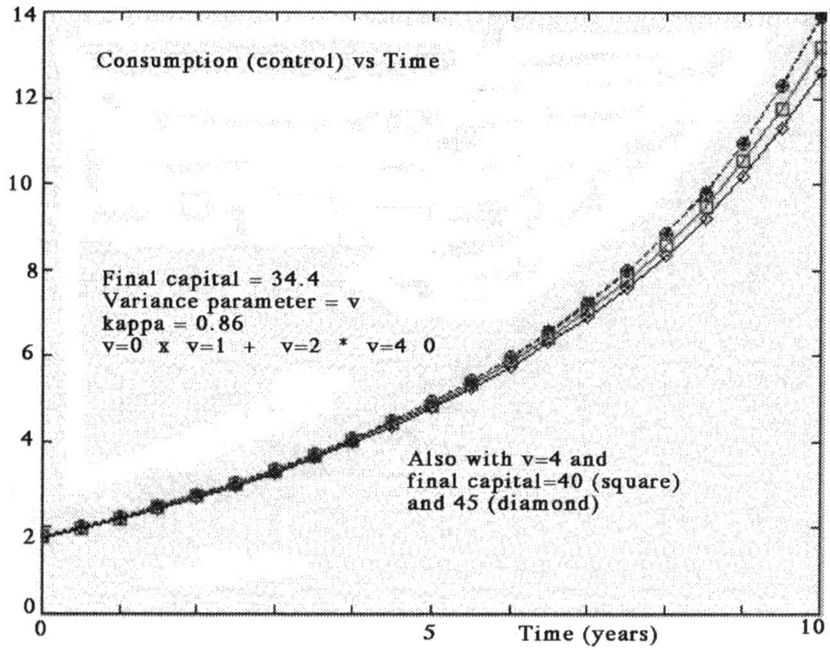

Consumption (control) vs Time
Final capital = 34.4
Variance parameter = v
kappa = 0.86
v=0 x v=1 + v=2 * v=4 0
Also with v=4 and
final capital=40 (square)
and 45 (diamond)
Time (years)
0
2
4
6
8
10
12
14
5
10

Fig. 6

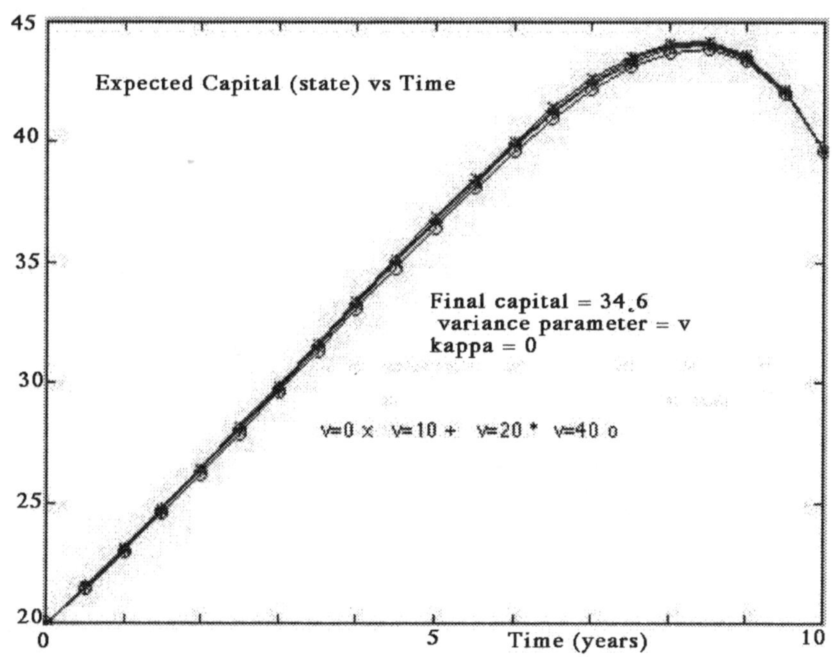

Expected Capital (state) vs Time
Final capital = 34.6
variance parameter = v
kappa = 0
v=0 x v=10 + v=20 * v=40 o
Time (years)
20
25
30
35
40
45
0
5
10

Fig. 7

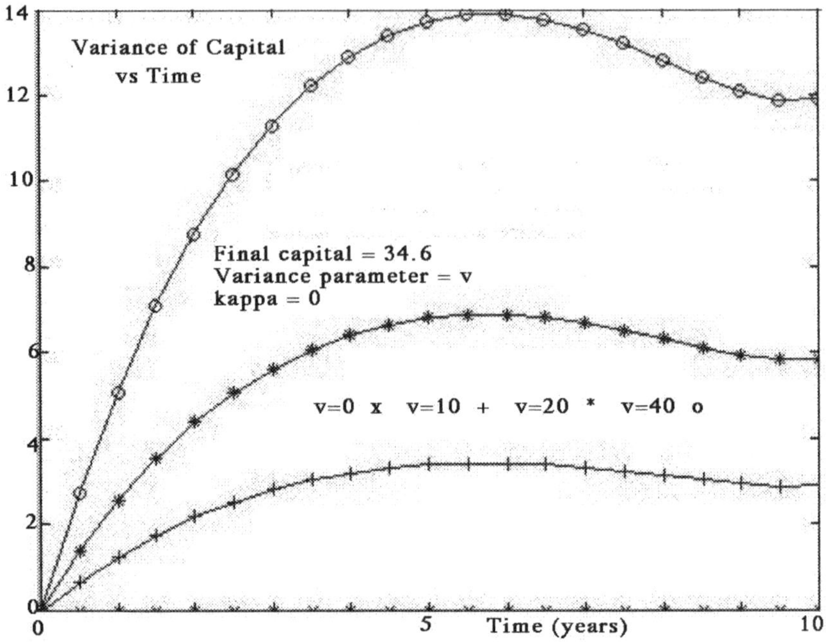
Variance of Capital
vs Time
Final capital = 34.6
Variance parameter = v
kappa = 0
v=0 x v=10 + v=20 * v=40 o
Time (years)
0
2
4
5
6
8
10
12
14

Fig. 8

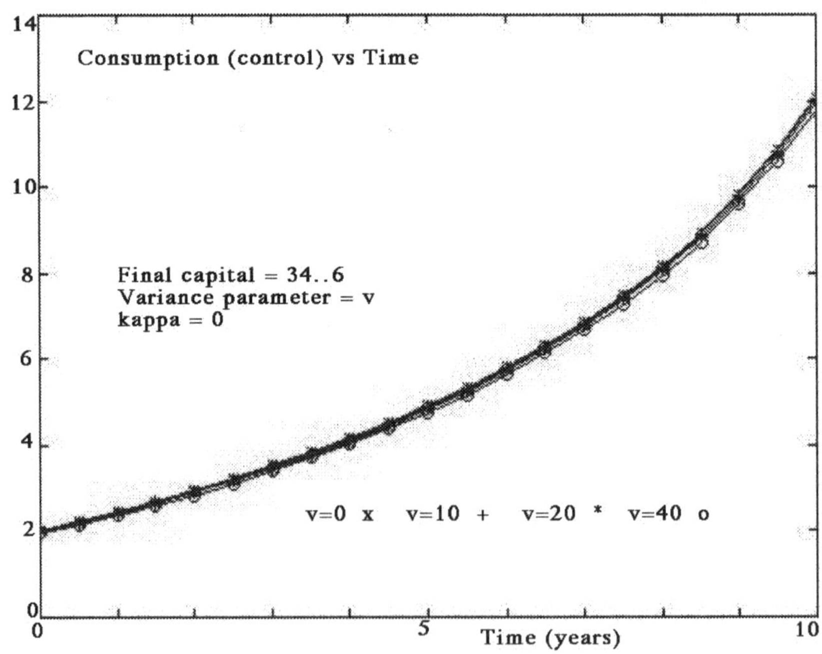
Consumption (control) vs Time
Final capital = 34..6
Variance parameter = v
kappa = 0
v=0 x v=10 + v=20 * v=40 o
Time (years)
0
2
4
5
6
8
10
12
14

Fig. 9

The code in *sys_dh* (for gradients of rhs of dynamic equations) becomes:

$$h_x = [0.842 * exp(0.02 * t) * 0.6 * x(1) \wedge (-0.4) - 0.05 - 0.0036;$$

$$\ldots - 2 * 0.00183 * x(2) - 2 * 0.00183 * x(1)]; \; h_u = [-1; 0];$$

Here $N/T = 2.0$; the bold term **2.0** is stated for $v = 1$, so is changed if the variance parameter v is changed.

The RIOTS_95 M-files for this stochastic growth model are as follows:

Calling program runKstoc.m
```
path(path, 'c:\ riots_95')
cd c:\ riots_95\ systems
! copy KendStoc\sys_*.m
clear simulate
N=20;
T=10;
u0=zeros(1,N+2-1);
t=[0:20/N:20];
x0=[15.0; 0.0];
format compact
[u,x,f]=riots(x0,u0,t,0.2,15.0,[],[400,.1,1],2)
```

Parameters function neq = sys_init(params)
```
if params == [],
neq = [1 2 ; 2 1 ; 8 1 ; 12 1];
else global sys_params
sys_params = params;
end
```

RHS of dynamic equations KendStoc\sys_h.m
```
function xdot = sys_h(neq,t,x,u)
global sys_params
xdot = [0.842*exp(0.02*t)*x(1)∧0.6-u(1)-0.05*x(1)-2.0*0.0018*x(2) ...
  8.0-2.0*(0.9354+0.00183*x(1))*x(2)];
% The term 8.0 above describes 2.0*variance parameter
% It is 0.0 for the deterministic model.
xdot = [0.842*exp(0.02*t)*x(1)∧0.6-u(1)-0.05*x(1)-2.0*0.0018*x(2) ...
  0.0-2.0*(0.9354+0.00183*x(1))*x(2)];
% The next line relates to the original Kendrick model
% xdot = [0.842*x(1)∧0.6 - u(1) - 0.05*x(1)];
```

Endpoint Constraint KendStoc\sys_g.m
```
function J = sys_g(neq,t,x0,xf)
global sys_params
```

```
F_NUM = neq(5);
if F_NUM == 1
J = 0;
elseif F_NUM == 2
J = xf(1) - 40.0;
end
```

Integrand of Objective Function KendStoc\sys_l.m

```
function z = sys_l(neq,t,x,u)
global sys_params
F_NUM = neq(5);
if F_NUM == 1
z = -10.0*(exp(-0.03*t)+0.86394)*u(1)^0.1;
  % The parameter 0.86394 gives weight to future
  % The following line describes the original Kendrick model
% z = -10.0*(exp(-0.03*t)+0.0)*u(1)^0.1;
else
z = -x(1) + 0.3;
end
```

Gradients for dynamic equation KendStoc\sys_dh.m

```
function [h_x,h_u] = sys_Dh(neq,t,x,u)
global sys_params
h_x = [0.842*exp(0.02*t)*0.6*x(1)^(-0.4)-0.05   -0.0036; ...
  -2.0*0.00183*x(2)   -2.0*0.00183*x(1)];
h_u = [-1   0];
```

Constraint Gradients KendStoc\sys_dg.m

```
function [J_x0,J_xf,J_t] = sys_Dg(neq,t,x0,xf)
global sys_params
F_NUM = neq(5);
J_x0 = [0 0 0]; J_xf = [0 0 0];
if F_NUM == 1
J_x0(1) = 0;
elseif F_NUM == 2
J_xf(1) = 1;
end
```

Gradients of Integrand KendStoc\sys_dl.m

```
function [l_x,l_u,l_t] = sys_Dl(neq,t,x,u)
global sys_params
l_x = 0;
l_u = -(exp(-0.03*t)+0.86394)*u(1)^(-0.9);
% The parameter 0.86394 gives weight to the future
% The following line describes the original Kendrick model
% l_u = -(exp(-0.03*t)+0.0)*u(1)^(-0.9);
```

Chapter 8

Optimization in Welfare Economics

8.1. Static and dynamic optimization

Optimal social welfare is generally defined with Pareto's mathematical definition of optimality of a vector objective function (see section 2.8). The analysis of social welfare and social choice (see Arrow and Reynaud, 1986; Arrow, Sen and Susumura, 2003; Schulz, 1988) also depend on vector optimization. Economic developments of these ideas are presented e.g in Sengupta and Fox (1969), Ehrgott (2000). In Islam (2001a), such ideas are applioed to decentralized and to centrally planned economies. Social welfare cannot always be measured by summing individual utilities, and a vector of utilities must be studied, with appropriate vector ordering (see e.g. Arrow and Scitovsski, 1969; Arrow and Intriligator, 1985; Sen, 1969). An economic model may maximize (in Pareto's sense) several competing welfare criteria or utilities $u_i(\cdot)$, or maximize some welfare function $W(u_1(\cdot), u_2(\cdot), \ldots)$, subject to constraints $g_i(x) \geq 0$, and perhaps also equality constraints. The classical economic assumptions of non-satiety of utility, and convexity of preference and production sets, are usually made here, but may be relaxed (see Schulz, 1988) for discussion of some non-classical models). The $u_i(\cdot)$ could be utilities of individual agents. While the welfare function is often taken as a weighted sum of individual utilities, the implied assumption that one agent's misfortune is compensated by another's good fortune is open to doubt. The usual assumption of convexity, under which a local optimun is a global optimum, is not always appropriate. When it fails, there may be multiple optima (see section 2.5). However, the methods of section 2, and such concepts as *invex,* relaxing *convex,* still apply.

An economic model may be *static,* thus not time-dependent, or it may be *dynamic,* thus evolving in time. Dynamic models' typically lead to optimal control formulations. For the basic formulations, refer to sections 1.2, 2.1, 2.8 and 2.13. Such models typically arise in optimal growth models, such as with allocation and social choice, sustainable resource management, intergenerational equity, and economic planning. They contain conflicting economic objectives, such as consumption utility maximization, achieving full employment, intergenerational economic equity, sustainability, and environmental requirements. The models considered include economic growth, decentralized planning, collective utility, and resource depletion. A multiobjective optimum consists of a set of Pareto points, among which a choice must eventually be made by other criteria (see Bolintineanu, 1993).

8.2. Some static welfare models

A number of models have been given in Islam and Craven (2001b), and are summarized here. The maximization of $W(u_1(\cdot), u_2(\cdot), \ldots)$, or the vector

(Pareto) maximization of a vector objective $\{u_1(\cdot), u_2(\cdot), \ldots),$, may be restricted by additional constraints, which put a floor below which no individual utility may fall. This adds extra terms to the Lagrangian condition (WKKT) for optimality, so that the optimum is no longer a Pareto maximum for the original model. The *non-satiety* property does not always apply. The utility for user i may be extended to:

$$u_i^{\#}(x) := u_i(x) + \sum_{j \neq i} \alpha_{ij} u_j(x),$$

to give some weight, depending on the coefficients α_{ij}, to the welfare of other users. Denote the vector of these new utilities by $MU(x)$, where $U(x) := \{u_1(x), u_2(x), \ldots, u_n(x))$, and M is a matrix made from the coefficients α_{ij}. The Pareto maximization of $MU(x)$ is equivalent to the Pareto maximization of $U(x)$ with respect to a different order cone:

$$Q = M^{-1}(\mathbf{R}_+^n);$$

see section 2.8.

A welfare function W may give a good description only in some central region, thus excluding outlying regions where some utility falls below a floor. A secondary objective function, that need not be only a function of the objective components, may be used to optimize over the set of Pareto optimum points (see Bolinteanu, 1993a, b), and this will determine a set of weights for the objective components. This may bear on the question of impossibility (Arrow, 1951; Arrow and Reynaud, 1996) or possibility (Sen, 1970).

In a *cooperative game* model described by Cesar (1994), agent i has his own objective function $J^i(.)$, and a weighted payoff function $\sum w_i J^i$ (x) is to be maximized, subject to the state vector x in a suitable feasible region X. This gives one of the Pareto maxima of the vector function:

$$J(.) := (J^1(,), J^2(.), \ldots)$$

over X, depending on the choice of weights w_i.

However, there are other game models to which invexity and related ideas may be applied. A zero-sum competitive game for two players leads to a *minimax* model, where there is a single objective function $J(u_1, u_2)$, which player 1 maximizes with respect to u_1, and player 2 minimizes with respect to u_2, subject to constraints $g(u_1) \leq 0$ and $k(u_2) \leq 0$. Define a Lagrangian function as:

$$L(u_1, u_2; v_1, v_2) := J(u_1, u_2) : -v_1 g(u_1) + v_2 k(u_2).$$

Under some (non-trivial) regularity assumptions (Craven, 1988), it is necessary for a minimax that the gradient of this Lagrangian with respect to (u_1, u_2) is zero, for some non-negative multipliers v_1, v_2, together with $v_1 g(x_1) =$

$0, v_2k(x_2) = 0$. These conditions are sufficient for a minimax under appropriate invexity assumptions (Craven, 1988).

A welfare model, described in Leonard and Long (1992) and Schulz (1988) has I individuals, G goods, R resources, and amount Y_r available of resource r. In terms of the state variables x_g^i = amount of good g used by individual i and y_{rg} = amount of resource r used to make good g, a welfare function:

$$W(U^1(x_1^1, \ldots, x_G^1), \ldots, U^I(x_1^I, \ldots\ , x_G^I\ )) \ \equiv W(U^1(\mathbf{x}^1), \ldots, U^I(\mathbf{x}^I)),$$

is maximized, subject to nonnegative variables and constraints:

$$a_i x_g^i = F_g(y_{1g}, \ldots\ , y_{Rg}\ ) \ \equiv F_g(\mathbf{y}_g)\ (g = 1, 2, \ldots, G)$$

$$\sum_g c_g y_{rg} = b_r\ (i = 1, 2, \ldots, R).$$

The Lagrangian is:

$$L(.) = W(U^1(\mathbf{x1}), \ldots, U^I(\mathbf{xI})) + \sum_g \pi_g[F_g(by^g) - \sum_i a_i x_g^i]$$

$$+ \sum_r \lambda_r[b_r - \sum_g c_g y_{rg}]$$

So necessary KKT conditions for a maximum are:

$$(\partial W/\partial U^i)(\partial U^i/\partial x_{gi}) - \pi_g a_i = 0 (i = 1, 2, \ldots, I).$$

If these conditions are sufficient, which happens if $-W(U^1(\cdot), \ldots, U^I(\cdot))$ and the constraint functions are invex with the same scale factor, then equilibrium conditions follow, thus under weaker assumptions than the usual]convex W and linear constraints. Otherwise, a *quasimax* of the welfare function is obtained, or a *vector quasimax* of the vector optimization problem.

A multiobjective optimum consists of a set of Pareto points, among which a choice must eentually be made by other criteria (see Bolintineanu, 1993). In the models discussed here, some changes in the model, such as adding "unselfish" utility terms, result in moving from one Pareto point of the original model to another. While each set of positive weights for the objectives determines one or more Pareto points, and each Pareto point determines one or more sets of weights, the weights are not an *a priori* choice by a decision maker, but may change during computation (see section 8.5). Moreover, a point satisfying the Karush-Kuhn-Tucker (KKT) conditions (or equivalent Pontryagin conditions for optimal control) is not necessarily a maximum.

8.3. Perturbations and stability

The extended welfare model of section 8.2 with utilities $u_i^{\#}(x)$ may be considered as Pareto maximizing the original utilities $u_i(x)$ with respect to a different (larger) order cone, and so the optima occur with some different values of the weights for the objectives, specified by the multiplier τ in WKKT.

A Pareto optimum p is *locally proper* (Craven, 1990) if the ratios of difference of objectives, $(u_j(x) - u_i(x))$, are bounded when x is in some neighbourhood of p. (For comparison, Geoffrion's *proper* Pareto maximum (Geoffrion 1968) considers all x.) Then (Craven, 1990) a Pareto maximum p is *locally proper* exactly when the multiplier $\tau > 0$, thus when each component $\tau_i > 0$. This dsescribes a sort of stability of the objective components near a proper Pareto maximum; there is a bound to how fast they change. There is an equivalent stability property also for *weak quasisaddlepoint* (see section 2.9), described in Craven (1990).

Consider now a linearization of the welfare model with utilities $u_i(x)$ around a Pareto maximum p, and approximate the extended welfare model. with α_{ij} replaced by $\epsilon\alpha_{ij}$, by a perturbation of this linearized model, with a small parameter ϵ. The model is then a *multi-objective linear program MLP*, including a parameter ϵ. For a chosen set of weights for the objective components, specified by τ, there results a linear program, with a small perturbation applied to the objective, but not to the constraints. A standard theory applies here. The point p is generally a vertex of the (polyhedral) fceasible region (satisfying the constraints), and p is not moved by the perturbation if ϵ is sufficiently small. Thus a standard welfare model will give some (small) weight to the welfare of other users.

8.4. Some multiobjective optimal control models

Craven and Islam (2003) give a multiobjective version of the *decentralized resource allocation model* of Arrow and Hurwicz (1960) (also Sengupta and Fox, 1969, page 62). If there are n producers, and now n objective functions, then a Pareto maximum of the vector function

$$\{f_1(x_1), f_2(x_2), \ldots, f_n(x_n)\}$$

may be considered, instead of a maximum of the sum of these components. The effect is to maximize $\sum \tau_i f_i(x)$ subject to constraints $x_i \in X_i, \sum g_i(x_i) \leq b$, with non-negative weights τ_i which sum to one. Different Pareto maxima arise from different choice of τ. If τ is fixed, and a convex problem is assumed, then there is a dual problem:

$$\text{MIN}_v \sum_i (m_i(v) + vb/n),$$

where v is a price attached to resource usage, and

$$m_i(v) = \text{MAX}(\tau_i f_i(x_i) - v g_i(x) \text{ subject to } (\forall i)\ x_i \in X_i.$$

However, the convex assumption may be reduced to *invex*; thus, if $x_i \in X_i \Leftrightarrow \xi_i(x) \leq 0$, and if all the functions $-f_i, \xi_i, g$ are invex with the same scale function.

Craven and Islam (2003) also presented an economic growth model involving two conmumption (control) variables, and one capital (state) variable $k(t)$. This model is now extended to r consumption variables $c^j(t)$. Motivated by the Kendrick-Taylor model (see section 4.2), consider the dynamic equation:

$$k(0) = k_0, \quad \dot{k}(t) = b(t)\xi(k(t)) - \sum c^i(t) \quad (0 \leq t \leq T),$$

and a vector objective to be maximized, with components:

$$F_i^i(k, c) = \int_0^T e^{-\rho t} U^i(k(t), c(t), t)dt - Q^i(k(T)),$$

with $c(t) := (c^1(t), \ldots, c^r(t))$ and utilities

$$U^i(k(t), c(t)) = c^i(t)^\sigma + \alpha \sum_{q \neq i} c^q(t)^\sigma,$$

in which (as in section 2) each user allows some weight to the welfare of other users. For this model, the Hamiltonian is:

$$\sum \tau_i \{e^{-\rho t} U^i(k(t), c(t), t)\} - \lambda(t)(\xi(k(t)) - \sum c^i(t)).$$

So the adjoint differential equation for the costate $\lambda(t)$ is:

$$-\dot{\lambda}(t) = \lambda(t)\xi'(k(t)), \quad \lambda(T) = \mu(k(T) - k^*),$$

with a weight μ and a constant k^* chosen so that a given terminal constraint on $k(T)$ is satisfied (see section 2.1). If the controls $c^i(t)$ are not subject to any active constraints, then Pontgryagin's principle requires that the gradient of the Hamiltonian with respect to the controls is zero. This gives:

$$e^{-\rho t} \tau_i^* \sigma (c^i(t))^{\sigma - 1} + \lambda(t) = 0,$$

where $\tau_i^* := \tau_i + \sum_{q \neq i} \alpha \tau_q$ is a modified multiplier for F_i. So here again the result of giving weight to the welfare of other users is to modify the multipliers attached to the objective components.

The behaviour is similar for a *collective utility* or *welfare* moldel, where the objectives to be maximized are:

$$\int_0^T e^{-rhot} \sum_q M_{iq} U((c^q(t))),$$

where $U(\cdot)$ is a utility function, $c^i(t)$ is consumption for user i,, $k(t)$ is capital stock, and thne matrix M of non-negative components M_{iq} describes the weighting each user gives to the others. This model may describe an altruistic social welfare planning model. As before, if there are no active constraints on the controls, then the multiplier vector τ when M is the identity matrix is shifted to $M^T\tau$.

8.5. Computation of multiobjective optima

If weights τ_i are chosen for the objectives multiobjective problem, then a scalar objectine $\sum \tau_i F_i(\cdot)$ may be optimized, by the methods usual for the constrained optimization of a single objective. For dynamic problems, differential equations must then be solved (see chapter 3.) But since the correct weights are seldom known *a priori*, it is usual to supply a *decision maker* (DM) with tradeoff information on the effects of changing the balance between the several objectives, so that the DM may prescribe a direction in which to change the weights. Often *goal programming* is used, this minimizing a weighted sum of deviations from goals, or targets (see e.g. Rustem, 1998). However, this weighting depends implicitly on the chosen scales of measurement for the several objectives.

Since Pareto optimal points are many, every opportunity should be taken to reduce the region to be searched. This may be done by placing bounds (e.g. floors) on some quantities, or on ratios of some objectives, where appropriate.

A secondary objective function may be optimized over the Pareto set (see Bolintineanu and Craven, 1992). This process is, to some extent, modelling the decision maker.

The choice of parameters, which may be varied in a secondary optimization, whether by a decision maker or by a computer model, may be critical to the understanding of a complicated multi-dimensional situation. If there are several objectives to be optimized, say $F_1(x)$, $F_2(x)$, $F_3(x)$ over some region E, then it may be appropriate to choose one objective, say $F_1(x)$, as of first importance, and then to maximize the single objective $F_1(x)$ over E, subject to additional inequality constraints, such as $F_2(x) \geq p_1$, $F_3(x) \geq p_2$, on the other objectives. There remains a secondary optimization over suitable values of the parameters p_1 and p_2. However, these are likely to have a more direct meaning, say in a businės situation, than would the tradeoffs between objectives.

All these details are relevant, both for computing a static problem, and for computing a dynamic problem. For a static problem, any of the standard codes for constrained nonlinear optimization (or for linear programming, if the problem is linear) may be used. For a dynamic problem, the evaluation of function values and gradients requires the solution of differential equations (for a problem in continuous time), or of difference equations (for a problem in discrete time). The computational aspects for a problem in continuous time are described in chapter 3.

Furthermore, for a dynamic (optimal control) model, there are several additional kinds of parametric constraints that may be appropriate. If some quantity $v(t)$ is to be constrained, then the possibilities include:

- a pointwise bound: $(\forall t)\ v(t) \geq p$;

• an average bound: $\int_0^T w(t)v(t)dt \geq p$ with some weighting function $w(t)$;
• in the case of stochastic optimal control (see chapter 7, and the portfolio models described in Sengupta and Fanchon, 1997; Tapiero, 1996, Ziemba and Vickson, 1975, for examples), some probabilistic bound may be appropriate, such as

$$\mathbf{Prob}\{(\forall t)v(t) \geq p\} \geq 0.9;$$

With some of these bounds (when behaviour at several times is combined), the Pontryagin principle does not apply; however KKT conditions are still applicable.

8.6. Some conditions for invexity

An optimal control problem has an equality constraint, the dynamic equation. Consider the problem (with a single objective F):

$$\text{MIN } F(z) \text{ subject to } G(z) \leq 0, K(z) = 0.$$

Let p be a KKT point, with multipliers λ and μ. Assume the following version (*Type I*) of invexity (Hanson and Mond, 1987):

$$(\forall z)F(z) - F(p) \geq F'(p_\eta(z,p); \quad -G(p) \geq G'(p)\eta(z,p); \quad 0 = K'(p)\eta(z,p).$$

If z satisfies the constraints, then:

$$F(z) - F(p) \geq F'(p)\eta(z,p) = -\lambda G'(p)\eta(z,p) - \mu K'(p)\eta(z,p) \geq \lambda G(p) + 0 = 0,$$

so a minimum is reached. If instead F is vector valued, then F is replaced by τF, and $\tau(F(z) - F(p)) \geq 0$ follows, proving that p is a weak minimum.

In an optimal control problem, z is replaced by the state and control (x, u), and p by a point $(\bar{x}, \bar{u})$ satisfying the Pontryagin conditions. If the problem is written (see section 2.2) as:

$$\text{MIN} F(x,u), \text{ subject to } \quad Dx = M(x,u), \quad G(x,u) \equiv G(u) \leq 0,$$

then the required invexity takes the form:

$$F(x,u) - F(\bar{x},\bar{u}) \geq F_x\alpha + G_u\beta, \quad G(\bar{u}) \geq G_u\beta, \quad 0 = ({\scriptstyle D}{+}M_x)\alpha + M_u\beta,$$

in which the two components $\alpha(x,u;\bar{x},\bar{u})$ and $\beta(x,u;\bar{x},\bar{u})$ replace $\eta(z,p)$, and F_x, etc. denote partial derivatives at $(\bar{x},\bar{u})$.

A composite function $f(g(.))$, where $g : \mathbf{R}^n \to \mathbf{R}^m_+$ and $f : \mathbf{R}^m_+ \to \mathbf{R}$ are differentiable, is invex at p if g is invex, f is convex at q (where $q = g(p)$)

and $f'(q)$ has positive components, since (Craven and Islam, 2003), setting $F(z) := f(q+z) - f(q), \quad g(x) - g(p) \geq g'(p)\eta(x,p)$, hence:

$$f(g(x)) - f(z(p)) = F(q + g(x) - g(p)) = F(q + g(x) - g(p)) - F(0)$$

$$\geq F'(0)(g(x) - g(p)) = f'(q)(g(x) - g(p) \geq f'(q)g'(p)\eta(x,p).$$

It is sometimes possible to replace an equality constraint by an inequality constraint. One example is an economic growth model, with a terminal constraint $k(T) = k_T$ on the capital $k(t)$. If the model maximizes some utility depending on consumption, then the terminal constraint may often be replaced by an inequality $k(T) \geq k_T$, since the maximization tends to reduce $k(T)$. In such a case, the sign of the Lagrange multiplier for this terminal constraint is already determined by the inequality.

8.7. Discussion

While welfare economic models are usually based on convex assumptions, some relaxation to generalized convexity is possible (see e.g. Schaible and Ziemba, 1981.) The less restrictive *invex* assumption allows various non-classical economic situations to be modelled. The *quasimax* concept provides a description of a stationary point, without specifying the Lagrange multipliers for the constraints. The discussion of shadow prices is related to a *quasidual* problem also not requiring convex assumptions.

The weights for objective components, in a vector optimization model, are not assigned *a priori*, but arise in the theory as multipliers. They are related to the stability of the model to perturbations, and they may change during a computation.

Chapter 9

Transversality Conditions for Infinite Horizon

9.1. Introduction

Economic and finance models for dynamic optimisation, expressed as optimal control over an infinite time horizon, have been much discussed by economists, since many issues in social choices have everlasting implications. However, there has been controversy about the form and validity of terminal costate conditions (*transversality conditions*) as time tends to infinity. In this chapter, such terminal conditions are proved: (i) when state and control tend to finite limits, and some likely assumptions about signs of gradients hold, (ii) when state and control tend to limits, with a restriction on the rate of convergence, and (iii) for growth models, when normalization by dividing by a growth factor leads to a model of type (ii). These cases cover many of the economic and finance models of interest. Some of these have definite social choice applications in the context of new[3] welfare economics. The analysis leads to a discussion of shadow prices, conditions sufficient for an optimum, and possible computational approaches for infinite-horizon models.

9.2. Critical literature survey and extensions

Economic models, especially for optimal growth and development, or for behaviour in time of a financial process, often lead to optimal control problems, over a finite time interval [0, T], or an infinite time interval $[0, \infty)$. Typically, the model is described by a *state function* $x(t)$ (usually vector-valued), describing e.g. capital accumulation, and a *control function* $u(t)$, describing .e.g. consumption. An objective, expressed as an integral involving the state and control, is optimized, subject to a dynamic differential equation, and constraints on the control. Under some standard assumptions, conditions necessary for a maximum or minimum of such a model over [0, T] consist of the following:

(i) the dynamic differential equation for a state function $x(t)$, with an initial condition for $x(0)$, and perhaps a terminal condition for $x(T)$;
(ii) an adjoint differential equation for a costate function $\lambda(t)$, with a terminal condition on $\lambda(T)$; and
(iii) Pontryagin's maximum (or minimum) principle, that a Hamiltonian function is optimized, with respect to the control function $u(t)$, at the optimal control.

Note that the standard assumptions include the restriction that any constraints on the control function apply to each time separately, thus excluding any restriction on the rate of change of the control, or any compensation between controls at different times. If (as with some economic models) the control is not constrained, then (iii) may be replaced by a weaker requirement, that the gradient of the Hamiltonian with respect to control is zero, and the assumptions are less stringent. The costate function, in an economic model, has the significance of a shadow price or a shadow cost.

The necessary conditions (i), (ii) and (iii) are not generally enough to imply an optimum. They become also sufficient conditions for a global optimum if the functions defining the control problem possess suitable *convex* properties, or (less restrictively) *invex* properties. This does not work if the terminal condition on $\lambda(T)$ is omitted, since the system then is not completely defined, except in the special case when the terminal state $x(T)$ is completely specified (with T fixed), for then $\lambda(T)$ is free. This terminal condition on $\lambda(T)$ has often been called a *transversality condition*, for historical reasons deriving from *time-optimal* problems, where the (variable) time T to reach a target is minimized. In that situation, the transversality condition appears as a description of how the optimal path crosses a *terminal line* (graphing state against T, as in Chiang (1992). Sethi and Thompson (2000) have presented $\lambda(T)$, for various cases when T is finite. (Some more general cases are also possible; see section 9.4.)

Transversality conditions have been much discussed (e.g. Leonard and Long, 1992; Burmeister and Dobell, 1970; Sengupta and Fanchon, 1997) for economic and financial models. Agénor and Montiel (1996) cited a modified transversality condition for infinite horizon (see section 4). Cuthbertson (1996) has reported another sort of transversality condition for a financial model with constant expected returns. However, a transversality condition is not significant on its own, but only as part of a complete set of necessary (and perhaps also sufficient) conditions. Judd (1998) discusses some economic models where the Euler equation (equivalent to the adjoint differential equation) without its terminal condition allows solutions which contradict economic requirements, so that a terminal condition is needed. (But it is already part of the necessary conditions). Moreover, the form of a transversality condition depends on features of the model, namely the terminal condition (if present) on the state function, and the endpoint term (if present) in the objective function. It is pertinent to consider how these two features affect such questions as stability of the model. Islam and Craven (2003b) have considered dynamic optimization models in finance, including cases of multiple steady states.

The role of the transversality conditions in economic and finance models has been summarized by Sengupta and Fanchon (1997, page 240), as:

"These conditions, often stated in terms of the adjoint (also called costate) variables, or the Lagrange multipliers associated with the dynamic equations of motion, seek to guarantee three things: uniqueness of the optimal trajectory, stability in the sense of convergence of the optimal path, and a condition of perfect foresight of the future."

It is remarked that uniqueness does not always occur for a non-convex model (Blanchard and Fischer, 1989), and that in the case of an infinite horizon

T (see section 3), some assumption about convergence of the optimal path is required to derive the transversality conditions. For such growth and finance models for an infinite horizon without a discount factor, the objective function is generally infinite. Following Ramsey (1928), this difficulty is avoided by considering a modified objective $\int_0^\infty [f(x(t), u(t)) - B\}dt$, where the *bliss point* B is a limiting value as time $t \to \infty$. But this modified objective will also diverge to an infinite value, unless the state and control are supposed to converge sufficiently fast (e.g. as $e^{-\beta t}$) to limiting values. Thus some assumption on convergence rates is already implicit in standard economic models. However, any analysis of uniqueness and stability for a dynamic model in economics and finance must include the correct transversality conditions.

Blitzer, Clarke and Taylor (1975) consider a model for finite horizon, whose objective is a utility function of consumption, plus an endpoint term depending on final capital. Several terminal conditions (at the finite horizon) are discussed, including a linear relation between final capital and consumption rate, and a lower bound for the final capital (thus attaching a value to capital after the horizon time). Versions of such conditions might also be applied to infinite-horizon problems.

Models with an infinite time domain $[0, \infty)$ are of economic significance. The standard Pontryagin theory does not immediately apply to an infinite time domain. Some authors (e.g. Leonard and Long, 1992) have applied the standard theory to a problem truncated to [0, T], with a terminal condition that $\lambda(T)$ equals the value of the costate at T for the infinite-domain problem. This obtains (ii) and (iii) for finite times t, but not a terminal condition on $\lambda(t)$ as t tends to infinity. Some authors have postulated such a condition (without adequate proof), since it is required to obtain sufficient conditions for an optimum.

Michel (1982) analysed an infinite-horizon model, with no terminal constraint on the state (trajectory), and obtained the transversality condition $\lambda(t) \to 0$ as $t \to \infty$ under some restrictions (non-negative objective integrand, optimal state contained in a suitable neighbourhood of feasible solutions). However, these (and other) assumptions "depend on properties of the optimal trajectory to be determined". Janin (1979) showed that $\lambda(t) \to 0$ at an exponential rate ($|\lambda(t)| \leq \text{const}\ e^{-\alpha t}$) for a class of growth models with infinite horizon, no terminal constraint on the state, a restriction (a coercivity constraint) on the dynamic equation, and a discount rate not too small. Although transversality does not always hold, the counter-examples of Halkin (1974) and Shell (1969) are (see Chiang, 1992) of an untypical character (the objective depends only on the terminal state and control). However, the derivation of transversality conditions for infinite horizon, including more general cases such as $\lambda(t)x(t) \to 0$, requires more analysis than "let T tend to infinity" in various authors. Michel (citing Janin, 1979) mentions requirements of "fast convergence"; sometimes these are implied by the assumption that an optimum is reached with a finite objective. In some financial models (see Cuthbertson, 1996), where an assumption that "expected returns are constant" replaces an optimization, a terminal condition is obtained, that a function of dividends tends to zero as time tends to infinity; this is also called "transversality".

Some economists have considered the case when the state and the control each tend to limits as t tends to infinity (see e.g. the discussion in Judd, 1998). But this does not generally happen for infinite-horizon models, especially if there are growth factors. The values of these limits need not be initially specified; an optimization calculation will find them. In this case, one may consider (see section 4) a *nonlinear transformation* of time t in $[0, \infty)$ to *scaled time* τ in [0,1], constructing τ by a suitable increasing function of t. Because of the assumed limits, $\tau = 1$ is in the domain of the problem (whereas $t = \infty$ is not). However, the transformed functions corresponding to m and f may become unboundedly large, as t tends to 1, upsetting the calculation. This situation is avoided if $x(t)$ and $u(t)$ converge sufficiently fast to their respective limits, as t tends to infinity. This leads to the expected (if not always proved) terminal conditions.

This chapter extends the literature in this area, by presenting three approaches to terminal (transversality) conditions for the costate function, for infinite horizon. These are listed as follows:

(a) There is a substantial class of infinite-horizon models, discussed in section 4, with the final state unrestricted, for which the terminal condition $\lambda(t) \to 0$ as $t \to \infty$ is a consequence of some assumptions about signs of gradients, and the assumed attainment (existence) of a finite optimum for the infinite horizon. However, the condition $\lambda(t)x(t) \to 0$ does not necessarily follow. An example is given where the state is unbounded, tending to no limit, and $\lambda(t) \to 0$ is not enough to determine a unique costate $\lambda(.)$. Applying the method of section 9.4 to problems where control and state tend to finite (unconstrained) limits, gives a large class of models where it is proved that the costate tends to zero, and this is the correct transversality condition to show sufficiency, under e.g. convexity or uniqueness assumptions.

(b) When terminal conditions on the state are imposed, involving the assumption that state and control tend to some limits as $t \to \infty$, then terminal conditions for $\lambda(t)$ as $t \to \infty$ are obtained in section 9.5, assuming that the convergence to these limits is not too slow (it must be fast enough that certain integrals have finite values).

(c) For some growth models, where limits are not reached, the state and control may be normalized (see section 9.6) by dividing by some growth factor (e.g. an exponential growth, or a power of t), so that case (b) applies to the normalized state and control. This approach solves the example cited in (a).

The economic interpretation of the costate $\lambda(t)$, and other multipliers which occur, in terms of shadow prices is discussed in section 9.7. Sufficient conditions for a (global) optimum, involving generalized convex (*invex*) assumptions, relaxing the usual *convex* assumptions, are discussed in section 9.8. Dynamic modelling under non-convex assumptions has been studied (see e.g.

Islam and Craven, 2004), but some questions remain to be answered.

The nonlinear time transformation has computational consequences (see section 9.9). If $x(t)$ and $u(t)$ tend to limits, then the functions in the problem (the right side of the dynamic equation, and the integrand of the objective function to be optimized) are changing only slowly when t is large. A description using *scaled time* τ can have the rates of change more uniform over τ in [0,1], than is possible over t in $[0, \infty)$. Consequently, a computation using *scaled time* may be practicable, perhaps only needing a moderate number of subintervals for the discretization, when a direct computation using original time t may be impracticable (see Craven, de Haas and Wettenhall, 1998).

Section 9.10 gives some discussion of special aspects of financial models.

9.3. Standard optimal control model for economic growth and financial dynamics[14]

Consider the following model, with a state function $x(t)$ (e.g. capital stock) and a control function $u(t)$ (e.g. consumption):

$$\text{MAX} \int_0^T e^{-\delta t} f(x(t), u(t))dt + \Phi(x(T))$$

subject to:

$$x(0) = x_0, \dot{x}(t) = m(x(t), u(t), t), q(x(T) = 0,$$

$$a(t) \leq u(t) \leq b(t) \ (0 \ \leq t \leq \ \text{T}).$$

Here T is finite and fixed; the endpoint constraint $q(x(T)) = 0$ is not always present; constraints on the control $u(t)$ are not always explicitly stated, although an implicit constraint $u(t) \geq 0$ is commonly assumed. If $q(.)$ or $\Phi(.)$ are absent from the model, they are replaced by zero.

Necessary Pontryagin conditions for a minimum of this model have been derived in many ways. In Craven (1995), the control problem is reformulated in mathematical programming form, in terms of a Lagrangian:

$$\int_0^T [e^{-\delta t} f(x(t), u(t)) + \lambda(t) m(x(t), u(t), t) - \lambda(t)\dot{x}(t) - \alpha(t)(a(t) - u(t))$$

$$-\beta(t)(u(t) - b(t) - \frac{1}{2}\mu[\Phi(x(t) - \mu^{-1}\rho]_+^2 - \frac{1}{2}\mu[q(x(T) - \mu^{-1}\nu]\delta(t-T)] \ dt.$$

with the costate $\lambda(t)$, and also $\alpha(t)$ and $\beta(t)$, representing Lagrange multipliers, μ a weighting constant, ρ and ν are Lagrange multipliers, and $\delta(t-T)$ is a Dirac delta-function. Here, the terminal constraint on the state, and the endpoint term $\Phi(x(T))$ in the objective, have been replaced by penalty cost terms in the

14 See also sections 2.2, 6.3.3, and 6.5.2.

integrand; the multipliers ρ and ν have meanings as shadow costs. (This has also computational significance – see section 8. The solution of a two-point boundary value problem, when $x(T)$ is constrained, has been replaced by a minimization.)

Then necessary Karush-Kuhn-Tucker (KKT) conditions follow. The adjoint differential equation (ii) follows from one part of KKT (the Lagrangian function has zero gradient with respect to x), integrating the term $\lambda(t)\dot{x}(t)$ by parts; an integrated term $\lambda(T)x(T)$ arises, leading to a terminal condition $\lambda(T) = 0$ in case $x(T)$ is free (thus, unconstrained). Other contributions to $\lambda(T)$ come from $q(.)$ and $\Phi(.)$, integrating the delta-function. The Pontryagin principle (iii) follows eventually, under some assumptions (such as bounded second derivatives of the functions f and g) from the other part of KKT (zero gradient of the Lagrangian with respect to x). Note that the proof requires a finite horizon T, and that the control constraints apply to each time t separately. The requirement of bounded second derivatives can be weakened to require that they are majorised by some function with finite integral.

The adjoint differential equation is obtained in the form:

$$-\dot{\lambda}(t) = e^{-\delta t} f_x(x(t), u(t)) + \lambda(t) m_x(x(t), u(t), t),$$

where f_x and m_x denote partial derivatives with respect to $x(t)$, together with a boundary condition (see Craven, 1995):

$$\lambda(T) = \Phi_x(x(T)) + \kappa q_x(x(T)),$$

in which Φ_x and q_x denote derivatives with respect to $x(T)$, and κ is a Lagrange multiplier, representing a shadow cost attached to the constraint $q(x(T)) = 0$. The value of κ is determined by the constraint that $q(x(T) = 0$. If $x(T)$ is *free*, thus with no terminal condition, and Φ is absent, then the boundary condition is $\lambda(T) = 0$. Note that $x(T)$ may be partly specified, e.g. by a linear constraint $\sigma^T x(T) = b$ (or $\geq$ b), describing perhaps an aggregated requirement for several kinds of capital. In that case, the terminal constraint differs from $\lambda(T) = 0$.

A diversity of terminal conditions for $\lambda(T)$ have been given in the economics literature (e.g. Sethi and Thompson, 2000); they are particular cases of the formula given above. For the constraint $q(x(T)) \geq 0$, the multiplier $\kappa \geq 0$.

9.4. Gradient conditions for transversality

There is a class of optimal control problems over infinite time, including some economic growth and dynamic finance models, where the boundary condition $\lambda(t) \to 0$ for costate follows without special assumption, supposing that an optimum is reached. Assume that the optimal control problem:

$$\text{MAX} \int_0^\infty f(x(t), u(t), t)dt \text{ subject to}$$

$$x(0) = 0, \dot{x}(t) = m(x(t), u(t), t), g(u(t)) \geq 0 \quad (0 \leq t < \infty),$$

where $f\ (.,.,.)$, $m(.,.,.)$ and $g(.)$ are differentiable functions, reaches an optimum at the point $(x(.), u(.)) = (x^*(.), u^*(.))$. (Usually this requires a discount factor $e^{-\delta t}$ included in f.) Then the truncated problem, for finite T:

$$x(0) = 0, \dot{x}(t)\ \mathrm{m(x(t),\ u(t),\ t)},\ g(u(t)) \geq 0\ (0\ \leq t < \infty),$$

reaches an optimum when $(x(t), u(t)) = (x^*(t), u^*(t))$ for $t < T$. Assuming that $u^*(.) > 0$, and that a constraint qualification, necessary Lagrangian conditions hold, in terms of a costate function $\lambda(.)$:

$$-\dot{\lambda}(t) = f_x(x^*(t), u^*(t), t) + \psi(t)m_x(x^*(t), u^*(t), t),$$

$$0 \geq -\theta(t)g_u(u(t)) = f_u(x^*(t), u^*(t), t) + \lambda(t)m_u(x^*(t), u^*(t), t)\ ,$$

assuming $g_u(u(t)) \geq 0$; the multiplier $\theta(t) \geq 0$. (Further hypotheses, such as bounded second derivatives of f and m, appropriate to proving Pontryagin's principle, are not needed here.)

Assume now that constants $b_1 > 0, b_2 > 0, b_3 > 0$ and $b_4 > 0$ exist, such that, for all $t \geq 0$:

$$f_x(x(t), u(t), t) \geq b_1, f_u(x(t), u(t), t) \geq b_2,$$

$$m_x(x(t), u(t), t) \geq b_3, m_u(x(t), u(t), t) \leq -b_4.$$

(Note that these are appropriate for an economic model in which the objective increases when the capital $x(t)$ and the consumption $u(t)$ each increase, and when the rate of growth of capital increases as existing capital increases, and decreases as consumption increases.) Assume that $(x(.), u(.))$ is in a neighbourhood N of $(x^*(.), u^*(.))$. For given $(x(.), u(.)$, denote $\rho(t) := \int_0^t m_x(x(t), u(t), t)dt$. From the equation for $\dot{\psi}(t)$,

$$(d/dt)(\lambda(t)e^{-\rho(t)}) = -f_x(x(t), u(t), t),$$

hence:

$$\lambda(t) = \psi(0)e^{-\rho(t)} - e^{-\rho(t)} \int_0^t f_x(x(s), u(s), s)e^{-\rho(s)}\ \mathrm{ds},$$

in which the integral *is* > 0 since $f_x(.) \geq b_1 > 0$, and $\rho(t) \geq b_3 t \rightarrow \infty$ as $t \rightarrow \infty$. So $\lambda(t) \leq \psi(0)e^{-\rho(t)} \rightarrow 0$ as $t \rightarrow \infty$. Also, if $\lambda(t) < 0$ for some t, then:

$$f_u(x^*t(t), u^*(t), t) + \psi(t)m_u(x^*(t), u^*(t), t)) \geq b_2 + (-\psi(t))b_4 > 0,$$

so cannot be ≤ 0. Hence $\lambda(t) \geq 0$ for all t. Hence, for all t:

$$0 \leq \lambda(t) \leq \lambda(0)e^{-\rho(t)} \rightarrow 0 \text{ as } t \rightarrow \infty.$$

Thus, for this class of optimal control model, the costate $\lambda(.)$ satisfies the terminal condition $\psi(t) \rightarrow 0$ as $t \rightarrow \infty$. It was not assumed that the state $x(t)$

tends to some limit as $t \to \infty$. Note that this does not work if an upper bound is imposed on $u(t)$.

However, a constraint $x(t) \geq b$ when t is sufficiently large could be adjoined. This is done by adding to the integrand $f(.)$ a penalty term of;

$$-\frac{1}{2}\mu[x(t) - b - \mu^{-1}\sigma(t)]_-^2,$$

where $[\cdot]_-$ means negative part, and $\sigma(t)$ is a non-negative multiplier. The result is to add a term $-\sigma(t)\chi_E(t)$ to the right side of the equation for $-\dot{\lambda}(t)$, and to $f_x(.)$ in the integral expression for $\psi(t)$, where E is the set of t where the bound $x(t) = b$ is reached. Thus $\dot{\lambda}(t) + \lambda(t)m(.) \geq -f_x(.)$, with equality for $t \notin E$. In this case, $\lambda(t)$ may tend to a limit $\lambda(\infty) > 0$ as $t \to \infty$.

If the above gradient conditions hold, and if also the state and control tend to finite (unconstrained) limits are $t \to \infty$, then the terminal condition $\lambda(t) \to 0$ as $t \to \infty$ is the precise requirement for a unique costate, and (under e.g. convexity or uniqueness assumptions) for sufficiency conditions for an optimum.

Consider now the example of a growth model:

$$\text{MAX} \int_0^\infty e^{-\delta t}[u(t)^\gamma + \zeta x(t)\}dt$$

$$\text{subject to } x(0) = x_0 > 0, \dot{x}(t) = \alpha x(t) - \text{ u(t) (t } \geq 0),$$

where $\delta > \alpha > 0$ (to ensure the objective is finite), $0 < \gamma < 1$ (e.g. $\gamma = \frac{1}{2}$). There are implicit constraints $x(.) > 0$ and $u(.) > 0$. The standard theory gives $-\dot{\lambda}(t) = -\zeta e^{-\delta t} + \alpha\lambda(t)$ and $-\gamma e^{-\delta t}u(t)^{\gamma-1} - \lambda(t) = 0$. Hence $\lambda(t) = (a + \kappa)e^{-\alpha t} - \kappa e^{-\delta t}$, where $a = \lambda(0)$ and $\kappa = \zeta/(\delta - \alpha)$, so $\lambda(t) \to 0$ as $t \to \infty$, without any assumption on $\lim_{t\to\infty}\lambda(t)$, but the parameter a is undetermined. However, the normalizing transformation $X(t) := e^{-\alpha t}x(t), U(t) := e^{-\alpha t}u(t), \Lambda(t) := e^{\alpha t}\lambda(t)$ leads to a unique normalized costate $\Lambda(t) = \Lambda(0) + \kappa(1 - e^{-(\delta-\alpha)t}) = -\kappa e^{-(\delta-\alpha)t}$ after applying the condition $\Lambda(t) \to 0$ as $t \to \infty$. (Hence $\lambda(t) = -\kappa e^{-\delta t}$.) Then $U(t)$ is determined by $\gamma e^{-(\delta-\alpha)t}U(t)^{\gamma-1} = -\Lambda(t) > 0$. This normalizing approach is pursued in section 9.6. Also, if the infinite horizon is replaced by a finite horizon T, then $\Lambda(t) = \kappa e^{-(\delta-\alpha)T} - e^{-(\delta-\alpha)t}$, which tends to the value for infinite horizon as $T \to \infty$. Thus this problem is stable to truncation of the infinite horizon to a large finite horizon T. (For a general discussion of such stability, see Craven (2003).) In contrast, a model in Judd (1998) is, in present notation:

$$\text{MAX} \int_0^\infty e^{-\delta t}U(u(t))dt \text{ subject to } x(0) = x_0, \dot{x}(t) = F(x(t) - \text{ u(t)},$$

assuming that $x(t)$ tends to a finite limit. But under what restrictions on U and F does such steady state exist? If, instead, the objective includes (as above) a state term, then growth models with no steady state can occur. Agénor and Montiel (1996) cite an infinite-horizon model with a transversality condition (in present notation) $e^{-\delta t}x(t) \to 0$ as $t \to \infty$, thus apparently not requiring the state to be bounded.

9.5. The model with infinite horizon

The model with finite T replaced by ∞ (and $\Phi(x(T))$ omitted), may be reduced to a control model of standard form by a a *nonlinear transformation* of time t in $[0, \infty)$ to *scaled time* τ in [0,1]. (A possible such transformation is given by $\tau = [e^{2\kappa t} - 1]/[e^{2\kappa t} + 1]$, with some scale parameter κ. Note that τ is an increasing function of t, with $\tau \to 1$ as $t \to \infty$.) Define $\hat{x}(\tau) = x(t)$, and $\hat{u}(\tau) = u(t)$. If $x(t)$ and $u(t)$ are assumed to tend to finite limits as $t \to \infty$, then $\hat{x}(\tau)$ and $\hat{u}(\tau)$ are defined (and continuous) at $\tau = 1$. Thus $\tau = 1$ is in the domain of definition for the problem (although $t = \infty$ is not). The result is a transformed problem, which is of the standard form for optimal control, except that some functions (such as the right hand side of the dynamic equation) may become unboundedly large as $\tau \to 1$. That situation is avoided if $x(t)$ and $u(t)$ are supposed to converge sufficiently fast to their limits, as $t \to \infty$.

Denote $x(\infty) := \lim_{t\to\infty} x(t), u(\infty) := \lim_{t\to\infty} u(t)$, and (if it exists) the limit $\lambda(\infty) := \lim_{t\to\infty}\lambda(t)$. Assume that, for some positive exponent β :

$$|x(t) - x(\infty)| \leq \text{ const } e^{-\beta t} \text{ and } |u(t) - u(\infty)| \leq \text{ const } e^{-\beta t}$$

Then (Craven, 2003) $|\hat{x}(t) - \hat{x}(1)|$ and $|\hat{u}(t) - \hat{u}(1)|$ are sufficiently small, as $t \to 1$, that the second derivatives of f and m, while perhaps unbounded, are majorised by a function with finite integral. The theory of section 9.3 then applies to the problem with scaled time, and then $\lambda(1) = \kappa q(x(1))$. Hence, for the given problem: $\lambda(\infty) = \kappa q(x(\infty))$. In particular, $q(x(T))$ might have a form $b^T x(T) - r$, so that a constraint $q(x(T)) = 0$ (or ≥ 0) would set a bound, or a lower bound, to the capital at the terminal time. It is not obvious whether an objective endpoint term $\Phi(x(\infty))$ would have meaning in this context.

9.6. Normalizing a growth model with infinite horizon

As stated in section 1, some different behaviours occur when the state does not tend to a steady state as $t \to \infty$, and may be unbounded. Consider then an optimal control problem, describing a *growth model* and financial dynamics:

$$\text{MAX } F(x, u) := \int_0^\infty e^{-\delta t} f(x(t), u(t))dt \text{ subject to:}$$

$$x(0) = x_0, \dot{x}(t) = m(x(t), u(t), t), \ a \ \leq u(t) \leq b \ (0 \ \leq t < \infty).$$

Instead of assuming that the state $x(t)$ and the control $u(.)$ tend to finite limits as time $t \to \infty$, assume instead that $x(t) = \mu(t)X(t)$, and $u(t) = \mu(t)U(t)$, where $\mu(.)$ is a *growth factor*, such as a power of t, and $X(t)$ and $U(t)$ each tend to limits as $t \to \infty$.

Consider in particular a dynamic equation:

$$\dot{x}(t) = ax(t)^{\beta} - u(t),$$

where $0 < \beta < 1$. If $u(.) = 0$ then $x(t)$ is proportional to $(c+t)^{\gamma}$, where c is constant, and $\gamma = 1/(1-\beta)$. Assume therefore $\mu(t) = (c+t)^{\gamma}$. Then:

$$\dot{X}(t) = (c+t)^{-1}[-\gamma X(t) + aX(t)^{\beta}\ ] \ - U(t).$$

Suppose that $U(t) \to 0$, and $X(t)$ is bounded. Then $\dot{X}(t) \to 0$, and $X(t)$ tends to a limit, as $t \to \infty$. If also $f(x(t), u(t))$ has the form $x(t)^{\xi}u(t)^{\zeta}$, then the integrand is $e^{-\delta t}X(t)^{\xi}U(t)^{\zeta}\mu(t)^{\xi+\zeta}$, and the integral is finite, under the above assumptions.

Consider now the further replacement of the coefficient a by $ae^{\rho t}$, thus introducing an exogenous growth factor into the dynamic equation. Let $\mu(t) = e^{-\sigma t}$, where $\sigma = \rho/(1-\beta)$. In this case, the bounds on the control are reduced to just $u(t) \geq 0$ for all t. Then

$$\dot{X}(t) = -\sigma X(t) + aX(t)^{\beta} - U(t).$$

The integrand becomes:

$$e^{-(\delta-\sigma\xi-\sigma\zeta)t}X(t)^{\xi}U(t)^{\zeta}$$

so the integral is finite provided that the discount rate:

$$\delta > \alpha(\xi+\zeta)/(1-\beta)$$

Integrating $x(t)^{-\beta}\dot{x}(t) = ae^{\rho t} - e^{\rho t z}(t)$, where $z(t) := e^{-\rho t}u(t)x(t)^{-\beta}$:

$$(1-\beta)x(t)^{1-\beta} = k_1 + a\rho^{-1}e^{\rho t} - z^*\rho^{-1}(e\rho t - 1) - \int_0^t e^{\rho s}[z(s) - z^*]ds,$$

assuming $z(t) \to z^* > 0$ as $t \to \infty, |z(s) - z^*| < k_2 e^{-\kappa t}(t \geq 0), 0 < \kappa < \rho$.

Then $e^{-\rho t}|\int_0^t e^{\rho s}[z(s) - z^*]ds| \leq (k_2/(\rho-\kappa))e^{-\rho t}[e^{(\rho-\kappa)t} - 1]$. Hence:

$$(1-\beta)x(t)^{1-\beta} = (a - z^*)\rho^{-1}e^{\rho t} + e^{-\rho t}r(t) \text{ where } r(t) \to 0 \text{ as } t \to \infty.$$

Thus $X(t) \to k_3$ as $t \to \infty$, with $\sigma = \rho(1-\beta)^{-1}$, $k_3^{1-\beta} = (a - z^*)\rho^{-1}$, since:

$$x(t) \sim k_3 e^{\sigma t}\ (t \to \infty).$$

Under these stated conditions, the convergence rate requirements of section 9.5 are satisfied for the reformulated problem:

$$\text{MAX} \int_0^\infty e^{-(\delta-\sigma\xi-\sigma\zeta)t} X(t)^\xi U(t)^\zeta dt \text{ subject to:}$$

$$X(0) = x_0,\ \dot{X}(t) = -\sigma X(t) + aX(t)^\beta - U(t),\ U(t) \geq 0\ (0 \leq t < \infty).$$

The corresponding adjoint differential equation is:

$$-\dot{\Lambda}(t) = \xi e^{-(\delta-\sigma\xi-\sigma\zeta)t} X(t)^{\xi-1} U(t)^\zeta + \Lambda(t)(-\sigma X(t) + aX(t)^\beta - U(t)),$$

where this new costate function $\Lambda(t)$ satisfies (from section 5) the terminal condition:

$$\Lambda(\infty) = \kappa q(X(\infty)).$$

Under these discussed conditions, the state $x(t)$ grows at an exponential rate, and a normalized state (divided by an exponential factor) tends to a finite limit. Agénor and Montiel (1996) discuss cases where this limit is zero.

9.7.1. Shadow prices

For a mathematical program with differentiable functions:

$$MAX_z\ F(z,q) \text{ subject to } G(z,q) \geq\ 0,\ K(z,q) = 0,$$

assume that an optimum is reached when $z = \bar{z}(q)$, when the perturbation parameter q is small. Denote by $V(q) := F(\bar{z}(q), q)$ the value of the perturbed objective function. Define the Lagrangian:

$$L(z;\rho,\sigma;q) := F(z,q) + \pi G(z,q) + \sigma K(z,q).$$

Under some standard assumptions (e.g. Craven 1995), the gradient:

$$V'(0) = L_q(\bar{z}(0); \bar{\pi}, \bar{\sigma}; 0),$$

where $\bar{\pi}$ and $\bar{\sigma}$ are the Lagrange multipliers corresponding to $q = 0$; they have the interpretation of *shadow prices.*

This may be applied to the optimal control problem in section 3, assuming that f and m also depend on a parameter q. Then:

$$V'(0) = \int_0^\infty [f_q(\bar{x}(t), \bar{u}(t), t, 0) + \bar{\lambda}(t) m_q(\bar{x}(t), \bar{u}(t), t, 0)$$
$$-\alpha(t) a_q(0) + \beta(t) b_q(0)] dt,$$

where $\bar{x}(t), \bar{u}(t), \bar{\lambda}(t)$ denote the unperturbed optimal solution.

In particular, if the only perturbation is of the dynamic equation, to:

$$\dot{x}(t) = m(x(t), u(t), t) + q(t),$$

then, if q is small:

$$V(q) - V(0) \approx \int_0^\infty \bar{\lambda}(t)q(t)dt.$$

Thus, $\bar{\lambda}(t)$ is not precisely a shadow price, but defines one by this integral. The model is only stable to such small perturbations q if they are such that this integral is finite. If the transversality condition is $\lambda(t) \to 0$ as $t \to \infty$, with $|\lambda(t)| \leq$ const $e^{-\alpha t}$, then $q(t)$ should be bounded. If instead $\lambda(t) \to \lambda^* \neq 0$ as $t \to \infty$, then qt) must tend to 0 sufficiently fast as $t \to \infty$, to make the integral converge. Thus the transversality condition, which itself depends on the terminal condition for the state, specifies the kind of perturbation of the dynamic equation for which the model is stable.

Introduce now an endpoint condition $x(T) \geq \zeta$, or $x(t) \to \zeta$ as $t \to \infty$ for an infinite horizon. Similarly to section 2, this adds a term of:

$$-\frac{1}{2}\mu[-x(t) + \zeta + q_2 + \mu^{-1}\rho]_+^2$$

to the integrand, where q_2 is a perturbation. The result (for $T = \infty$) is:

$$V(q, q_2) - \text{V(0,0)} \approx \int_0^\infty \bar{\lambda}(t)q(t)dt + \rho q_2 .$$

The Lagrange multiplier ρ is determined by the constraint that $x(T) = \zeta$, or $x(t) \to \zeta$ as $t \to \infty$.

9.8. Sufficiency conditions

Karush-Kuhn-Tucker (KKT) conditions (or equivalent Pontryagin conditions), including the boundary condition for costate, are generally necessary, but not sufficient, for an optimum. Thus:

Optimum exists	$\Rightarrow$ KKT conditions	$\Rightarrow$? Optimum is reached,
(& some regularity)	(including boundary conditions)	

where $\Rightarrow$? means "implies, but only under further conditions to be stated". These further conditions could be:

(I): An optimum is otherwise known to exist, and a unique solution exists to KKT. Then this solution is the optimum.

or (II): the Hamiltonian $f(.) + \psi m(.)$ is concave in $(x(.), u(.))$ (see e.g. Leonard and Long, 1992). But, since in general the components of $\lambda(t)$, which relates to an equality constraint, can take either sign (see e.g. Leonard and Long, 1992) this criterion is seldom applicable, unless $m(.)$ is linear in $x(.)$ and $u(.)$. (Arrow (1970) seems to assume that $\lambda(t) \geq 0$).

(III): Invex hypotheses hold (convex is a special case) – see Craven, 1995; Islam and Craven, 2003).

The constraints of the problem:

$$\text{MAX } F(z) \text{ subject to } G(z) \leq 0,\ K(z) = 0,$$

may be combined as $Q(z) \in$ -S, where S is a convex cone; then "invex at z^*" requires that, for all z and some scale function $\eta(.,.)$:

$$-F(z) + F(z^*) \geq (-F'(z^*))\eta(z, z^*);$$

$$Q(z) - Q(z^*) - Q'(z^*)\eta(z, z^*) \in S.$$

Hence $G(z) - G(z^*) \geq G'(z^*)\eta(z.z^*)$ and $K(z) - K(z^*) = K'(z^*)\eta(z, z^*)$.

For the convex case, where $\eta(z.z^*) = z - z^*, K$ must then be linear. Since $K(z) = 0 = K(z^*)$, it suffices to have $0 = \Theta K'(z^*)\eta(z, z^*)$, where Θ is the Lagrange multiplier.

Consider now a control problem with $z = (x(.), u(.))$, and $m_z = (m_x, m_u)$. Since:

$$\Theta K(z) = \int_0^T \lambda(t)[-\dot{x}(t) + m(x(t), u(t), t)]dt$$

$$= \int_0^T \{\dot{\lambda}(t)x(t) + m(x(t), u(t), t)\}dt,$$

$$\Theta K'(z)\xi = \int_0^T [\dot{\lambda}(t) + \lambda(t)m_z]\xi(t)dt = \int_{0T} \lambda(t)[-\dot{\xi}(t) + m_z\xi(t)]dt,$$

here using the boundary condition $\lambda(T) = 0$ (or $x(T)$ fixed) in two integrations by parts. Hence $\xi(t) := \eta((x(t), u(t)), (x^*(t), u^*(t))$, for the considered $(x(t), u(t))$, must satisfy the differential equation:

$$-\dot{\xi}(t) + m_z(x^*(t), u^*(t), t)\xi(t).$$

It is sometimes possible to transform the variables from $(x(.), u(.))$ to other variables $(X(.), U(.))$, so as to produce a linear differential equation for $\dot{X}(t)$. This happens e.g. for the Kendrick-Taylor model for economic growth, for which the transformed objective also is concave (see Islam and Craven, 2003a).

9.9. Computational approaches for infinite horizon models

A optimal control model in economics and finance may have an infinite time horizon to describe the future interest of society. Several methods have been proposed for computing infinite horizon models in economics and finance so that the results show the future interest of society for an infinite time period. They include the following (Chakravarty, 1969; Islam, 2001a; Islam and Craven, 2003a):

• Including infinite time issues by a salvage value term in the objective function or by terminal constraints or transversality conditions (see Chichilnisky, 1996 ; Islam and Craven, 2003c).
• Computation of models with conditions so that each finite horizon result coincides with the relevant segment of an infinite horizon model. (Note that this gives theoretical results, but to compute one would also need some endpoint condition for the costate.)
• Computation of steady state equations, with a transversality condition at a truncated horizon. (Note that this assumes that the state and control tend to steady states, as e.g. in Judd (1998). This approach does not deal with the approach to the steady state.)
• Computation of models with sufficiently long horizon so any increase in time after that will not seriously change the optimal solution. (This assumes that the infinite-horizon optimum is well approximated by a sufficiently long finite horizon. But this is sometimes the theoretical point in question.)
• Computation of a bang-bang model which can generate steady state results within finite time periods.

The first three methods for specifying the terminal constraint (transversality condition) in a finite-horizon computational model are those usually considered for infinite-horizon models in economics and finance.

The nonlinear time transformation described in section 4, when applicable, gives a further computational method. The control problem with infinite horizon is reduced to a standard control problem on a finite time domain [0,1]. Given the assumption of (sufficiently) "fast convergence rate", the transformed problem has well-behaved functions for the integrand of the objective, and the right side of the dynamic equation.

For the models usually considered, the dynamic equation for the state is stable to small perturbations (as may arise in computation) when computed with time increasing from zero; and then the adjoint equation is only stable when computed with time decreasing from the horizon, which requires a transversality condition as initial point.

Consider the dynamic equation:

$$x(0) = x_0, \dot{x}(t) = m(x(t), u(t), t) \ (\mathrm{t} \ \geq \ 0),$$

with $m(.,.,.)$ continuous. For a given $u(.)$, write $\dot{x}(t) = \varphi(x(t), t)(t \geq 0)$. Assume that $x(t)$ tends to a finite limit $\bar{x}$ as $t \to \infty$, then (from continuity) $\dot{x}(t) \to 0$ as $t \to \infty$; then assume also that:

$$|\varphi(x(t), t)| \leq \omega(t), \text{ where } \int_0^\infty \omega(t)dt \text{ is finite.}$$

Change the time scale (as in section 5) by $t = \sigma(\tau), \tau = \rho(t) := \sigma^{-1}$ (t), where $\sigma(.)$ is increasing differentiable, $\sigma(0) = 0, \sigma(1) = \infty$. Let $X(\tau) := x(\sigma(\tau))$.

Then:

$$(d/d\tau)X(\tau) = \varphi(X(\tau), \sigma(\tau))\sigma'(\tau), X(0) = x_0 (0 \le \tau \le 1).$$

In order to solve this differential equation numerically, the right side needs to be bounded as $\tau \to 1$, otherwise the computation will be inaccurate, and perhaps unstable. The right side is bounded if $\omega(\sigma(\tau))\sigma'(\tau)$ is bounded, or equivalently if $\omega(t)/\rho'(t)$ is bounded as $t \to \infty$. In particular, if $\omega(t) = ae^{-\alpha t}$ for positive a, α, then the bounded requirement is fulfilled if $\rho'(t) \ge \alpha e^{-\alpha t}$, thus if $\rho(t) = 1 - e^{-\alpha t}$, satisfying $\rho(0) = 0$ and $\rho(t) \to 1$ as $t \to \infty$. Then $\sigma(\tau) = -\alpha^{-1} ln(1-\tau)$, so $\sigma'(\tau) = (1-\tau)^{-1}$, hence $\sigma'(\rho(t)) = (1-(1-e^{-\alpha t}))^{-1} = e^{\alpha t} = a/\omega(t)$.

However, this construction assumes that $x(t)$ tends to a limit, so will not apply to a *growth model,* unless the variables are changed as in section 9.5.

The adjoint differential equation for the control problem is:

$$-\dot{\lambda}(t) = e^{-\delta t} f_x(x(t), u(t)) + \lambda(t) m_x(x(t), u(t), t),$$

with a boundary condition for $\lambda(\infty)$. Assume the growth restrictions (for $a_1, a_2 > 0$) :

$$|m(x(t), u(t), t)| \le a_1 \omega(t), |m_x(x(t), u(t), t)| \le a_2 \omega(t), \int_0^\infty \omega(t) dt < \infty.$$

This must hold when $(x(.), u(.))$ is in a neighbourhood of the optimal solution. Let $\Lambda(\tau) := \lambda(\sigma(\tau))$ and $U(\tau) := u(\sigma(\tau))$. Then the adjoint differential equation transforms to:

$$-(d/d\tau)\Lambda(\tau) = [\Lambda(\tau) m_x(X(\tau), U(\tau), \sigma(\tau))\sigma'(\tau)$$

$$+ e^{-\delta\sigma(\tau)} f_x(X(\tau), U(\tau)]\sigma'(\tau).$$

Here the assumption with $\omega(.)$ ensure that the coefficient of $\Lambda(\tau)$ is bounded as $\tau \to 1$.

Consequently, one of the usual computing packages for optimal control may be used, and a moderate number of subdivisions of scaled time may suffice, instead of a large number, when errors may accumulate. In case (a), when the state tends to a limit (or, more generally, remains bounded), and the growth restrictions hold, then a scaled computation can cover the whole time horizon, up to infinity. In case (b), the state is unbounded, and then a truncated interval [0, T] for large T, with some endpoint term at T, can be mapped to scaled time [0,1] for computation. (For an example of such an optimal control computation with scaled time and a large horizon T, see Craven, de Haas and Wettenhall (1998).) However, when the state and control can be normalised, as in section 6, to new variables that tend to finite limits, then case (a) may be applied.

The growth model from section 9.4, with a known solution, may illustrate the computational issues involved. If $e^{-\alpha t}x(t)$ and $e^{-\alpha t}u(t)$ are taken as new state and control functions, thus dividing by a growth factor $e^{\alpha t}$, then the problem becomes:

$$\text{MAX} \int_0^\infty [e^{-(\delta-\alpha\gamma)t}u(t)^\gamma + e^{-(\delta-\alpha)t}x(t)]dt$$

$$\text{subject to:} \quad x(0) = x_0, \dot{x}(t) = -u(t) \ (t \ \geq 0).$$

The optimum is $u(t) = \alpha b e^{-\alpha t}, x(t) = x_0 -$ b(1 $-e^{-\alpha t}$), where b is constant. The time transformation $t = -\sigma^{-1}\ln(1-\tau), dt/d\tau = (1-\tau)^{-1}$, with $X(\tau) = x(t)$ and $U(\tau)(1-\tau) = u(t)$, converts the problem to the computable form:

$$\text{MAX} \int_0^1 [(1-\tau)^{-1+\gamma+(\delta-\alpha)/\sigma}U(\tau)^\gamma + \zeta(1-\tau)^{-1+\delta/\sigma}X(\tau)]d\tau,$$

$$X(0) = x_0, (d/d\tau)X(\tau) = U(\tau) \ (0 \ \leq \tau \leq 1.$$

9.10. Optimal control models in finance: special considerations

While the transversality conditions derived above apply also to financial models, some special features may apply to financial models. Financial systems are arguably less stable, because a speculative aspect enters. It is not so obvious that an infinite horizon is so relevant to a financial model. It is perhaps relevant to consider a periodic model, where the horizon T is finite (and perhaps not too large), and there are end conditions at time T, which must provide initial conditions for a re-run of the system for a further interval of length T. An economist's assumption that capital has no value after the horizon T cannot apply here.

9.11. Conclusions

It has been argued that transversality conditions are needed in dynamic economic and finance models. Their role and validity have been extensively discussed for infinite-horizon economic and finance models. This chapter has been able to show that the models in the existing literature are seriously restricted, generally assuming steady states (though a growth model need not have one), and requiring unstated assumptions about the convergence rate. The models in the literature have been extended, and transversality conditions for infinite horizon have been established for a large class of applicable models, including some growth models. A mathematical approach has been established, suggesting methods for computation over an infinite horizon. The extensions have provided an improved framework for treating the transversality conditions in optimal growth, development, and financial models.

Chapter 10
Conclusions

The objective of tbis book is to develop optimization models for economics and finance, which can be applied to characterize and make social choices about the optimal welfare state of the economy, and to other uses as discussed in section 1.2, on the basis of the principles of new[3] welfare economics. These types of optimization models, based on welfare economics, are appropriate, since they allow an explicit incorporation of social value judgments and the characteristics of the underlying socio-economic organization in economic and finance models, and provide realistic welfare maximizing optimal resource allocation and social choices, and decisions consistent with the reality of the economy under study. For this purpose, a variety of models have bveen developed in different chapters of the book, including the following types of models:

- models for optimal growth and development,
- small stochastic perturbations),
- finance and financial investment models (and the interaction between financial and production variables),
- modelling sustainability over long time horizons,
- boundary (transversality) conditions, and
- models with several conflicting objectives.

Both analytic and computational models concerning the aggregate and multi-agent economy have been studied in the preceding chapters. They relate to different economic and financial conditions, including deterministic macroeconomic dynamics, optimal economic growth and development, financial dynamics, sustainable growth, models with some uncertainty, cooperative games, and infinite-horizon programming. These questions have been approached using mathematical and computational techniques for optimization, including some extensions to current methods for optimization. The topics include the following:

- when is an optimum reached, and when it it unique?,
- relaxation of the conventional convex (or concave) assumptions on an economic or financial model,
- associated mathematical concepts such as *invex* (relaxing *convex*) and *quasimax* (relaxing *maximum*), and the circumstances when these apply,
- multiobjective optimal control models, and the related Pontryagin theory,
- related computational methods and programs,
- sensitivity and stability,
- discontinuous (jump) behaviours,
- optimization with an infinite horizon, related to the rate of approach to

a steady state.

To show the potential of these approaches, this book has emphasized algorithms and computations, including a new MATLAB package called SCOM, for optimal control.

The implications of the results from these models, concerning resource allocations that maximize welfare, have been briefly discussed. They show the plausibility of this optimization approach, and give some support to its validity. The mathematical extensions to traditional methods for economic and financial modelling offer scope for further research. The different chapters have offered modelling approaches to many aspects of optimal social choice, forecasting using optimal models, optimal policy choice, market simulation, and planning. These have potential for future research and practical applications.

Bibliography

Ackerman F., Kiron, D. and Goodwin, N.R. (1997). *Human Well-Being and Economic Goals,* Island Press, Washington.

Agénor, P.-R. and Montiel, P. J., (1996). *Development Macroeconomics,* Princeton University Press, Princeton, N. J.

Amir, R., (1997). A new Look at Optimal Growth under Uncertainty, *Journal of Economic Dynamics and Control,* 22, 61-86.

Amman, H., Kendrick, D. and Rust, J. (eds.), (1996). *Handbook of Computational Economics,*Elsevier, Amsterdam.

Aoki, M. ,(1989). *Optimization of Stochastic Systems,* Academic Press, San Diego.

Aronsson, T. Johansson, P. and Lofgren K., (1997). *Welfare Measurement, Sustainability and Green National Accounting: A Growth Theoretical Approach, New Horizons in Environmental Economic Series*, Edward Elgar, Cheltenham.

Arrow, K.J., (1951). Alternative Approaches to the Theory of Choice Under Risk-Taking Situations,*Econometrica,* vol. 29, no. 4, pp. 404-431.

Arrow, K.J., (1971). *Essays in the Theory of Risk-Bearing,* North Holland, Amsterdam.

Arrow, K. and Hurwicz, L. (1960). Decentralization and competition in resource allocation, in *Essays in Economics and Econometrics* , University of North Carolina Press.

Arrow, K. J. and Intriligator, M. D. (eds.), (1981). *Handbook of Mathematical Economics* , vol. 1, North Holland, Amsterdam.

Arrow, K.J. and Intriligator, M.D. (eds.), (1986). *Handbook of Mathematical Economics,* vol. III, North Holland, Amsterdam.

Arrow, K. J. and Kurz, M., (1970). *Public Investment, the Rate of Return, and Optimal Fiscal Policy,* Johns Hopkins Press, Baltimore.

Arrow, K. and Reynaud, H., (1996).*Social Choice in Multiobjective Decision Making*, M.I.T. Press, Cambridge, Mass.

Arrow, K. J., Sen, A., and Suzumura (eds.) K., (2003). *Handbook of Social Choice and Welfare Economics,* North Holland, Amsterdam.

Barro, R. J. and Sala-i-Martin, X., (1995). *Economic Growth*, McGraw-Hill, New York

Blanchard, O. J. and Fischer, S.,(1989). *Lectures on Macroeconomics,* MIT Press, Cambridge, Mass.

Blitzer, C. R., Clark, P. B., and Taylor, L., (1975). *Economy-Wide Models and Development Planning,* Oxford University Press.

Bolintineanu, S., (1993a). Optimality conditions for minimization over the (weakly or properly) efficient set, Journal of Mathematical Analysis and Applications 173 (1993b), 523-541.

Bolintineanu, S., (1993). Necessary conditions for minimization over the (weakly or properly) efficient set, Journal of Mathematical Analysis and Applications 173, 523-541.

Bolintineanu, S. and Craven, B. D., (1992). Multicriteria sensitivity and shadow costs, Optimization 26 , 115-127.

Borda, J.C.,, (1953). Memoire sur les Elections au Scrutin, Histoire de l'Academie Royale des Sciences, Translated by Alfred de Grazia, 'Mathematical Derivation of an Election System', Isis, vol. 44, no. 1-2, pp. 42-51

Bos, D., Rose, M. and Seidl, C. (eds.), 1988, *Welfare and Efficiency in Public Economics,* Springer Verlag, Heidelberg.

Brekke, K.A., (1997). *Economic Growth and the Environment: On the Measurement of Income and Welfare*, Edward Elgar, Cheltenham

Brooke, A., Kendrick, D., Meeraus, A. and Raman, R. (1997). *GAMS: A User's Guide*, GAMS Home Page.

Brandimarte, P., (2002). *Numerical Methods in Finance : a MATLAB-based introduction,* Wiley, New York.

Broadway, R. and Bruce, N., (1984). *Welfare Economics,* Basil Blackwell, Oxford.

Brock, W., (1971). Sensitivity of optimal growth paths with respect to a change in target stocks, *Zeitschrift für National Ökonomie , Supplementum 1*, 73-89.

Burmeister, E. and Dobell, A., (1970). *Mathematical Theories of Economic Growth*, Macmillan, London.

Campbell, J. Lo,, and MacKinlay A., (1997). *The Econometrics of Financial Markets* , Princeton University Press, Princeton, N.J.

Campbell, J. J. and Viceira, L. M., (2002). *Strategic Asset Allocation: Portfolio Choices for long term investors,* Oxford University Pross, New York

Cesar, S., (1994). *Control and Game Models of the Greenhouse Effect*, Springer Verlag, Heidelberg.

Chakravarty, S. (1969). *Capital and Development Planning*, MIT Press, Cambridge.

Chiang, A. C., (1992). *Methods of Dynamic Optimization,* Waveland Press, Prospect Heights.

Chichilnisky, G., (1977). Economic development and efficiency criteria in the satisfaction of basic needs, *Applied Mathematical Modeling,* 1.

Chichilnisky, G., (1996). An axiomatic approach to sustainable development, *Social Choice and Welfare*, 13 (2), 219-248.

Clarke, C. and Islam, S. M. N., (2004). *Economic Growth and Social Welfare;* Series: *Contributions to Economic Analysis,* North-Holland Publishing Co., Amsterdam.

Craven, B. D., (1977). Lagrangean conditions and quasiduality, Bulletin of the Australian Mathematical Society vol. 16, pp. 325-339.

Craven,B. D., (1978). *Mathematical Programming and Control Theory* Chapman & Hall, London.

Craven, B. D., (1981). Duality for generalized convex fractional programs, in Schaible, S. and Ziemba, W. T. (eds.), *Generalized Concavity in Optimization and Economics*, Academic Press, New York, pp. 473-490.

Craven, B. D., (1988). Lagrangian conditions for a minimax, *Proceedings of Conference on Functional Analysis/Optimization* , Australian National University, Canberra, pp. 24-33.

Craven, B. D., (1989). A modified Wolfe dual for weak vector minimization, Numerical Functional Analysis and Optimization 10 (9 & 10), 899-907.

Craven, B. D., (1990). Quasimin and Quasisaddlepoint for Vector Optimisation, Numerical Functional Analysis and Optimization , 11, 45–54.

Craven, B.D., (1994). Convergence of discrete approximations for constrained minimization, *Journal of the Australian Mathematical Society, Series A,* **35** , 1-12.

Craven, B. D., (1995). *Control and Optimization*, Chapman & Hall, London .

Craven, B. D., (1998). Optimal control and invexity, *Computers and Mathematics with Applications* 35 (5), 17-25.

Craven, B. D., (1999a). Multicriteria optimal control, *Asia-Pacific Journal of Operational Research* 16 (1), 53-62.

Craven, B. D., (1999b). Optimal control for an obstruction problem, *Journal of Optimization Theory and Applications* 100 , 435-439.

Craven, B. D., (2001). On nonconvex optimization, *Acta Mathematica Vietnamica* 26 (3) 249-256.

Craven, B. D., (2002). Optimal control of an economic model with a small stochastic term, (submitted for publication).

Craven, B. D., (2002b). Global invexity and duality in mathematical programming, *Asia-Pacific Journal of Operational Research* 19, 169-175.

Craven, B. D., (2003). Optimal control on an infinite domain, (submitted for publication).

Craven, B. D., de Haas, K., and Wettenhall,J. M., (1998). Computing Optimal Control, Dynamics of Continuous, Discrete and Impulsive Systems 4, 601-615.

Craven B. D., Glover B. M., and Lu, Do Van 1996. Strengthened invex and perturbation theory, Zeitschrift für Operations Research (Mathematical Methods of Operations Research) 43, 319-326.

Craven, B. and Islam, S. M. N., (2001). Computing Optimal Control on MATLAB : The SCOM Package and Economic Growth Models, in *Optimisation and Related Topics* , A. Rubinov et al., eds., Volume 47 in the Series *Applied Optimization* , Kluwer Academic Publishers.

Craven, B. and Islam, S. M. N. (2003). On multiobjective optimal control models for growth, planning, finance and sustainabily, (submitted for publication).

Craven, B. D. and Luu, D. V., (2000). Perturbing convex multiobjective programs, *Optimization* 48, 391-407.

Cuthbertson, K., (1996). *Quantitative Financial Economics : Stocks, Bonds, and Foreign Exchange* , John Wiley, Chichester, England.

Davis, B. E. and Elzinga, D. J., (1972). The solution of an optimal control problem in financial modeling, *Operations Research* 19, 1419-1473.

Diewert, W. E., (1981). Generalized Concavity and Economics, in *Generalized Concavity in Optimization and Economics* , S. Schaible and W. T. Ziemba (eds.), Academic Press, New York.

Dutta, P. K. , (1993). On specifying the parameters of a development plan, in *Capital, Investment and Development* , K. Basu, M. Majumdar and T. Mitra (eds.), Blackwell, Oxford, pages 75-98.

Ehrgott, M., (2000). *Multicriteria Optimization*, Springer, Berlin.

Faucheux, S., Pearce, D. and Proops, J. (eds.), (1996). *Models of Sustainable Development*, Edward Elgar, Hants, UK.

Fiacco A. V. and McCormick G.P., (1968). *Nonlinear Programming: Sequential Unconstrained Minimization Techniques*, Wiley, New York.

Fleming, W. M. and Rishel, R. W., (1975). *Deterministic and Stochastic Optimal Control*, Springer-Verlag, Berlin.

Forrester, J. W., (1971). *World Dynamics*, Wright Allen Press, Cambridge, Mass.

Fox, K., Sengupta J. and Thorbecke, E., (1973). *Theory of Quantitative Economic Policy*, North Holland, Amsterdam.

Gass, S. and Hemos, C., (2001). *Encyclopedia of Operations Research and Management Science,* Kluwer Academic Publications, London.

Goh, C.J. and Teo, T.L. (1987). *MISER, an Optimal Control Software*, Department of Industrial and Systems Engineering, National University of Singapore.

Geoffrion, A. M., (1968). Proper efficiency and the theory of vector maximization, *Journal of Mathematical Analysis and Applications* 22,618-630.

Gourieroux. C. and Janiak, J., (2001). *Financial Econometrics* , Princeton University Press, Princeton.

Haavelmo, T., (1954). Studies in the Theory of Economic Evolution, North-Holland Publishing Co., Amsterdam.

Hakansson, N., (1975). Optimal Investment and Consumption Strategies under Risk for a Class of Utility Functions, in Ziemba, W. T. and Vickson, R. G., *Stochastic Optimisation Models in Finance* , Academic Press, New York.

Halkin, H., (1974). Necessary conditions for optimal problems with infinite horizons, Econometrica, 42, 267-272.

Hanson, M. A., (1980). On the sufficiency of the Kuhn-Tucker conditions, *Journal of Mathematical Analysis and Applications* 80, 545-550.

Hanson, M. A. and Mond, B. M., (1987). Necessary and sufficient conditions in constrained optimization, *Mathematical Programming* 87, 51-58.

Hausman, D. and McPherson, M., (1996). *Economic Analysis and Moral Philosophy,* Cambridge University Press, Cambridge.

Heal, G.M., (1973). *The Theory of Economic Planning*, North-Holland, Amsterdam.

Heal, G., (1998). *Valuing the Future: Economic Theory and Sustainability* , Columbia University Press, New York.

Intriligator, M.D., (1971). *Mathematical Optimization and Economic Theory,* Prentice-Hall, New Jersey.

Islam, S. M. N., (1999). Bayesian Learning in Optimal Growth under Uncertainty, in T. Brenner (ed.), *Computational Techniques for Modelling Learning in Economics,* pp. 283-304, Advances in Computational Economics, vol. 11,

Kluwer Academic Publishers, London.

Islam, S. M. N., (2001a). *Optimal Growth Economics : An Investigation of the Contemporary Issues, and Sustainability Implications,* North Holland Publishing, Series - Contributions to Economic Analysis, Amsterdam.

Islam, S. M. N., (2001b). *Applied Welfare Economics: Measurement and Analysis of Social Welfare by Econometric Consumption Models,* CSES Research Monograph 1/2001, Melbourne.

Islam S. M. N. and Craven B. D., (2001a). Computation of non-linear continuous optimal growth models: experiments with optimal control algorithms and computer programs, *Economic Modelling : The International Journal of Theoretical and Applied Papers on Economic Modelling,* 18, 551-586.

Islam, S. M. N. and Craven, B. D., (2001b). Static Vector Optimization in Welfare Economics: Some Extensions, Journal of Multi-Criteria Decision Analysis (to appear).

Islam, S. M. N. and Craven, B. D., (2003a). Computation of non-linear continuous optimal growth models. (submitted for publication).

Islam, S. and Craven, B. D., (2003b). Dynamic optimization models in finance: some extensions to the framework, models and computation. (submitted for publication).

Islam, S. M. N. and Craven, B. D., (2003c). *Models and Measurement of Sustainable Growth and Social Welfare*, in Social Responsibility: Corporate Governance Issues, Batten, J. A. and Fetherston, T. A., eds., pages 223-252, JAI - Elsevier Science, Amsterdam.

Islam, S. and Craven, B. D., (2004). Some extensions of nonconvex economic modelling: invexity, quasimax, and new stability conditions, Journal of Optimization Theory and Applications (to appear).

Islam, S. M. N., and Oh, K. B., (2001). E-Commerce Finance in the Knowledge Economy: Issues, Macroeconomic Determinants and Public Policies, Fourth Annual Conference on Money and Finance in the Indian Economy, 13-15 Dec. Mumbai.

Islam, S. M. N. and Oh, K. B., (2003). *Applied Financial Econometrics in E-Commerce,* North-Holland Publishing Co, Amsterdam.

Islam, S.M.N. and Watanapalachaikul, S. (2004). *Empirical Finance: Modelling and Analysis of Emerging Financial and Stock Markets,* (in the Physica series *Contributions to Economics,*) Springer, Heidelberg.

Janin, R., (1979). Conditions nécessaires d'optimalité dans un problème d'optimalité en horizon infini, *Comptes Rendus de l'Academie des Sciences de Paris,* A289, 651-653.

Jennings,L., Fisher, M. E., Teo, K. L. and Goh, C. J., (1998). *MISER 3.2, an Optimal Control Software*, Dept. of Mathematics and Statistics, University of Western Australia, Perth.

Judd, K.L., (1998). *Numerical Methods in Economics,* MIT Press, Cambridge.

Karatzas, J., (1997), Lectures on the Mathematics of Finance, American Mathematical Society, Providence.

Karatzas, J. and Shreve, K. S.l, (1998). Methods of Mathematical Finance, Springer, New York.

Keller, E. and Sengupta, J., (1974). Sensitivity Analysis for Optimal and Feedback Controls Applied to Growth Models, *Stochastics*, vol. 1, pp. 239-266.

Kelly, D.L. and Kolstad, C.D., (1997). Bayesian Learning, Growth, and Pollution, Department of Economics, University of California, Santa Barbara.

Kendrick, D.A. and Taylor, L., (1971). *Numerical Methods and Nonlinear Optimizing Models for Economic Planning*, in H. B. Chenery (ed.) *Studies in Development Planning,* Harvard University Press, Cambridge, Mass.

Kendrick, D., (1981). *Stochastic Control for Economic Models*, McGraw Hill Book Company, New York.

Kendrick, D., (1987). Software for Economic Optimal Control Models, in Carraro, C. Sartore, D. *Developments of Control Theory for Economic Analysis*, Kluwer Publishers, Dordrecht.

Kendrick, D.A., (1990). *Models for Analysing Comparative Advantage*, Kluwer Academic, Dordrecht.

Kendrick, D.A. and Taylor, L., (1971). Numerical Methods and Nonlinear Optimizing Models for Economic Planning, in H. B. Chenery ed. Studies in Development Planning, Harvard University Press, Cambridge, Mass.

Kurz, M., (1968). Optimal economic growth and wealth effects, *International Economic Review* 9, 348-357.

Laffont, J-J., (1988). *Fundamentals of Public Economics*, MIT Press, Cambridge, Mass.

Land Economics, (1997). Special Issue on Sustainability.

Leonard D. and Long N. V., (1992). *Optimal Control Theory and Static Optimization in Economics,* Cambridge University Press, Cambridge, U.K.

Li, J.X., (1993). *Essays in Mathematical Economics and Economic Theory*, Ph.D. Dissertation, Department of Mathematics, Cornell University, Ithaca.

Li, J.X., (1998). Numerical Analysis of a Nonlinear Operator Equation Arising from a Monetary Model, *Journal of Economic Dynamics and Control,* 22, 1335-1351.

Lucas, R., (1978). Asset Prices in an Exchange Economy, *Econometrica,* **46**, 1429-45.

Malliaris, A.G. and Brock, W.A., (1982). *Stochastic Methods in Economics and Finance,* Elsevier Science, Amsterdam.

Mangasarian, O. L., (1969). *Nonlinear Programming*, McGraw-Hill, New York.

Markowitz, H., (1959). *Portfolio Selection: Efficient Diversification of Investments* , Wiley, New York.

MATLAB: The Language of Technical Computing, (1997). The MathWorks, Natick.

Michel, P., (1982). On the transversality condition in infinite horizon optimal problems, *Econometrica* 50(4), 975-985.

Mitra, T., (1983). Sensitivity of optimal programmes with respect to changes in target stocks : the case of irreversible investment, *Journal of Economic Theory* 29, 172-184.

Mitra, T. and Ray, D., (1984). Dynamic optimization on a non-convex feasible set : some general results for non-smooth technologies, *Zeitschrift für National Ökonomie* **44** 151-175.

Mufti, I., (1970). *Computational Methods in Optimal Control Problems,* Lecture Notes in Operations Research and Mathematical Systems, Springer Verlag, Heidelberg.

Radner, R., (1982). Equilibrium Under Uncertainty, in K.J. Arrow, and M.D. Intriligator (eds) *Handbook of Mathematical Economics,* Vol. 1, Ch 20, North-Holland Publishing Company, Amsterdam.

Ramon, M. and Scott , A., Ed., (1999). *Computational Methods for the Study of Dynamic Economies,* Oxford University Press, Oxford.

Ramsey, F.P., (1928). A Mathematical Theory of Saving, *The Economic Journal,* 543-59.

Rawls, J., (1972). *A Theory of Justice*, Clarendon, Oxford.

Rustem, B., (1998). *Algorithms for Nonlinear Programming and Multiple-Objective Decisions* , Wiley, Chichester.

Schulz N., (1988). Welfare economics and the vector maximum problem, in *Multicriteria Optimization in Engineering and in the Sciences* , W. Stadler (ed.), Plenum Press, New York and London.

Schaible, S. and Ziemba, W. T. (eds.), (1981). *Generalized Concavity in Optimization and Economics* , Academic Press, New York

Schwartz, A., (1996). Theory and Implementation of Numerical Methods Based on Runge-Kutta Integration for Solving Optimal Control Problems, Ph.D. thesis, Department of Electrical Engineering and Computer Sciences, University of California, Berkeley.

Schwartz,A., Polak, E. and Chen, Y., (1997). *Recursive Integration Optimal Trajectory Solver 95: A Matlab Toolbox for Solving Optimal Control Problems, (Version 1.0 for Windows)*, Stanford University, California.

Sen, A., (1997). *Choice, Welfare and Measurement*, Harvard University Press, Cambridge.

Sen, A. (1999). The possibility of social choice, *American Economic Review* 89 (3), 349-378.

Sengupta, J.K. and Fanchon, P., (1997). *Control Theory Methods in Economics*, Kluwer Academic, Boston.

Sengupta, J.K. and Fox, K.A., (1969). *Economic Analysis and Operation Research : Optimization Techniques in Quantitative Economic Models*, North Holland, Amsterdam.

Sengupta, J. K. and Fox, Karl A., (1969). *Optimization Techniques in Quantitative Economic Models* , North-Holland, Amsterdam.

Sethi, S, and Gerald L. Thompson, (2000). *Optimal Control Theory : Applications to Management Science and Economics,* Kluwer Academic Publishers, Boston.

Shell, K., (1969). Applications of Pontryagin's maximum principle to economics, in H. W. Kuhn and D. P. Szëgo, eds., *Mathematical Systems Theory and Economics,* Vol. 1, Springer Verlag, New York, pp. 273-275.

Smulders, J.A., (1994). *Growth, Market Structure, and the Environment,* PhD thesis, Tillburg University, Tillburg.

Stevens, J., (1993). *The Economics of Collective Choice,* Westview Press, Oxford.

Tabak, D., and Kuo, B. C., (1971). *Optimal Control by Mathematical Programming*, Prentice-Hall,New Jersey.

Takayama, A., (1985). *Mathematical Economics*, Cambridge University Press, Cambridge.

Tapiero, C.S., (1998). *Applied Stochastic Models and Control for Insurance and Finance*, Kluwer Academic, London.

Taylor, J. B. and Uhlig, H., (1990). Solving Nonlinear Stochastic Growth Models: A Comparison of Alternative Solution Methods, *Journal of Business and Economic Statistics,* 8, 1-17.

Teo, K.L., Goh, C.S. and Wong, K.D., (1991). *Unified Computational Approach to Optimal Control Problems*, Longman Scientific and Technical.

Thompson G. L et. al., (1993). *Computational Economics: Economic Modeling with Optimization Software* (Chapters 21 to 22 and Appendices A&B), Scientific Press.

Tintner, G. and Sengupta, J.K., (1969). *Stochastic Economics: Stochastic Processes Control and Programming,* Academic Press, New York.

Vickson, R. G., and Ziemba W.T., (1975). *Stochastic Optimisation Models in Finance,* Academic Press, New York.

Weitzman, M. C., (1976). On the welfare significance of national product of a dynamic economy, *Quarterly Journal of Economics*, 20, 156-162.

Zahedi, F. M., (2001). Group Decision Making, *Encyclopedia of Operations Research and Management Science,*, S.I. Gass and C. M. Harris (eds.), Kluwer Academic Publishers,London, pages 265-271.

Zenios, S.A., (ed.), (1993). *Financial Optimization,* Cambridge University Press, Cambridge.

Ziemba, W. T. and Vickson, R. G. (eds.), (1975). *Stochastic Optimization Models in Finance*, Academic Press, New York.

Index

Dynamic Modeling and Econometrics in Economics and Finance

1. P. Rothman: *Nonlinear Time Series Analysis of Economic and Financial Data*. 1999
ISBN 0-7923-8379-6
2. D.M. Patterson and Richard A. Ashley: *A Nonlinear Time Series Workshop*. 1999
ISBN 0-7923-8674-4
3. A. Mele and F. Fornari: *Stochastic Volatility in Financial Markets*. 2000
ISBN 0-7923-7842-3
4. I. Klein and S. Mittnik (eds.): *Contributions to Modern Econometrics*. 2002
ISBN 1-4020-7334-8
5. C. Le Van and R.-A. Dana: *Dynamic Programming in Economics*. 2003
ISBN 1-4020-7409-3
6. Eugenie M.J.H. Hol: *Empirical Studies on Volatility in International Stock Markets*. 2003
ISBN 1-4020-7519-7
7. B.D. Craven and S.M.N. Islam (eds.): *Optimization in Economics and Finance*. Some Advances in Non-Linear, Dynamic, Multi-Criteria and Stochastic Models. 2005
ISBN 0-387-24279-1

FSC
www.fsc.org
MIX
Papier | Fördert gute Waldnutzung
FSC® C083411

Kolibri 360 GmbH
Ferdinand-Jühlke-Straße 7
99095 Erfurt, Deutschland
produktsicherheit@kolibri360.de